MIRABILIS
A CARNIVAL OF CRYPTOZOOLOGY AND UNNATURAL HISTORY

DR. KARL P.N. SHUKER

Anomalist Books
*San Antonio * Charlottesville*

Mirabilis: A Carnival of Cryptozoology and Unnatural History
Copyright © 2013 by Karl P.N. Shuker
ISBN: 978-1-938398-05-6

All rights reserved, including the right to reproduce this book or portions thereof in any form whatsoever.

Book design by Seale Studios

For information, go to AnomalistBooks.com, or write to:
Anomalist Books, 5150 Broadway #108, San Antonio, TX 78209

Some of the chapters or sections of chapters in this book are reprinted, expanded, updated, or amalgamated versions of the following previously published articles or *ShukerNature* blog posts:

"Antiquarian's Menagerie," *Fortean Times*, No. 279 (September 2011): 58-59.
"The Megalopedus, the Sukotyro, and a Very Cryptic Cabinet of Curiosities," *ShukerNature*, http://www.karlshuker.blogspot.co.uk/2012/12/the-megalopedus-sukotyro-and-very.html, 13 December 2012.
"Titanic Sea Turtles—More Than Just a Maritime Myth?" *Practical Reptile Keeping*, May 2012: 36-39.
"Giant Tortoises of the Lost and Found," *Practical Reptile Keeping*, June 2012: 36-40.
"Mysterious Accounts of Megaturtles," *Practical Reptile Keeping*, July 2012: 36-39.
"Horned Hares and Jackalopes," *Small Furry Pets*, No. 1 (Winter 2011/12): 40-42.
"Along Came a Spider...," *Practical Reptile Keeping*, March 2012: 38-41.
"The Hobbit Actor and a Real-Life Shelob—Or Something Pretty Darn Close?" *ShukerNature*, http://www.karlshuker.blogspot.co.uk/2011/07/hobbit-actor-and-real-life-shelob-or.html, 16 July 2011.
"Forgotten Cryptids, Part One," *Paranormal*, No. 55 (January 2011): 50-55.
"Forgotten Cryptids, Part Two," *Paranormal*, No. 56 (February 2011): 32-37.
"Never Tangle With a Tygomelia—Or Tango With a Tokandia!" *ShukerNature*, http://www.karlshuker.blogspot.co.uk/2011/02/never-tangle-with-tygomelia-or-tango.html, 24 February 2011.
"Steller's Sea-Bear—A Polar Bear in Japan?" *ShukerNature*, http://www.karlshuker.blogspot.co.uk/2011/01/stellers-sea-bear-polar-bear-in-japan.html, 5 January 2011.
"A Sea Snail With Antlers...and Paws!" *ShukerNature*, http://www.karlshuker.blogspot.co.uk/2012/04/sea-snail-with-antlersand-paws.html, 16 April 2012.
"Zebro—An Equine Mystery From Iberia," *Flying Snake*, 1 (No. 1; April 2011): 46-47.
"The Tailed Slow Lorises of Assam's Lushai Hills—An Enduring Primatological Mystery," *ShukerNature*, http://www.karlshuker.blogspot.co.uk/2013/02/the-tailed-slow-lorises-of-lushai-hills.html, 24 February 2013.

"Going Wild Over the Wild American Hound—Your Assistance is Requested!" *ShukerNature*, http://www.karlshuker.blogspot.co.uk/2011/04/going-wild-over-wild-american-hound.html, 13 April 2011.

"Sea Dragons, Fairy Loaves and Serpents of Stone—Fables and Fossils of Lyme Regis," *The Dorsetarian*, http://www.darkdorset.co.uk/the_dorsetarian/0/fossil_folklore, 24 July 2010.

"Madagascar's Elusive Mega-Lemurs and Mini-Men," *ShukerNature*, http://karlshuker.blogspot.co.uk/2010/02/madagascars-elusive-mega-lemurs-and.html, 17 February 2010.

"Catch a Kilopilopitsofy!" *Fortean Times*, No. 131 (February 2000): 48.

"More Madagascan Mystery Mammals," *ShukerNature*, http://karlshuker.blogspot.co.uk/2010/03/more-madagascan-mystery-mammals.html, 17 March 2010.

"Between a Roc and a Hard Place," *ShukerNature*, http://karlshuker.blogspot.co.uk/2010/12/between-roc-and-hard-place.html, 29 December 2010.

"You've Been Trunkoed!" *Fortean Times*, No. 275 (May 2011): 42-47.

"The Trunko Probability," *Fortean Times*, No. 279 (September 2011): 72.

"The Entombed Toads Enigma," *Paranormal*, No. 54 (December 2010): 32-35.

"Headless and Petrified," *Fortean Times*, No. 103 (October 1997): 50.

"Hunting Africa's Crocodiles," *Practical Reptile Keeping*, August 2012: 36-39.

"Crocodile-Frogs and Crocodile-Snakes?" *Practical Reptile Keeping*, September 2012: 36-39.

"Maritime Mysteries of the Crocodilian Kind," *Practical Reptile Keeping*, October 2012: 36-40.

"The Giant Beaver: A Legend Come To Life?" *Small Furry Pets*, No. 4 (Autumn 2012): 52-55.

"The Curious Case of Hans Schomburgk's Missing Micro-Squirrels," *Small Furry Pets*, No. 7 (April-May 2013): 52-55.

"Flower-Generated Birds, and Turtles That Fly," *ShukerNature*, http://karlshuker.blogspot.co.uk/2012/06/flower-generated-birds-and-turtles-that.html, 16 June 2012.

"Sonnerat's Non-Existent Penguins (and Kookaburra) of New Guinea," *ShukerNature*, http://www.karlshuker.blogspot.co.uk/2013/05/sonnerats-non-existent-penguins-and.html, 1 May 2013.

"The Gorgakh, and Two Lesser-Known Indonesian Cryptids," *ShukerNature*, http://karlshuker.blogspot.co.uk/2010/02/the-gorgakh-and-two-lesser-known.html, 9 February 2010.

Karl Shuker has sought permission for the use of all illustrations and substantial quotes known by him to be still in copyright. Any omission brought to his attention will be rectified in future editions of this book.

CONTENTS

Introduction 1

Chapter 1: **The Megalopedus, the Sukotyro, and a Very Cryptic Cabinet of Curiosities** 3

Chapter 2: **Mega-Turtles and Titanic Tortoises—In the Wake of Giant Crypto-Chelonians** 11

Chapter 3: **Horned Hares and Jackalopes—An (Un)Natural History** 29

Chapter 4: **Along Came a (Giant) Spider...** 35

Chapter 5: **Never Tangle with a Tygomelia—Seeking Forgotten Mystery Beasts from Bygone Times** 49

Chapter 6: **Sea Dragons, Fairy Loaves, and Serpents of Stone—Fables and Fossils of Lyme Regis** 63

Chapter 7: **Giant Lemurs and Dwarf Hippos—On the Track of Madagascan Mystery Beasts** 71

Chapter 8: **You Know When You've Been Trunkoed—It's the Surreal Thing!** 93

Chapter 9: **Toads in Holes and Frogs in Throats—The Enigma of Entombed Life** 111

Chapter 10: **Crocodilian Monsters and Mysteries** 125

Chapter 11: **Mega-Beavers and Micro-Squirrels—Two Extremes in Crypto-Rodentology** 141

Chapter 12: **Flower-Generated Birds and Turtle-Headed Eels—Extracting the Ordinary from the Extraordinary in Cryptozoology** 153

Bibliography 165

Acknowledgements 173

About the Author 175

Index 181

In memory of my dear mother,
Mary Doreen Shuker
(January 29 1921 - April 1 2013)

Whatever good there may be in me came from you.
Thank you for blessing my life by being in it
as my mother.
You were, are, and always will be quite simply
the best person I shall ever know,
and I love you with all of my heart.
God bless you, little Mom,
please wait for me,
watch over me in this lonely existence of mine now,
and come for me when my time here is over.
Au revoir, Mom, until we meet again.

INTRODUCTION

Ex umbris et imaginibus in veritatem.
("Out of shadows and phantasms into the truth.")
— On the headstone of the Venerable John Henry Cardinal Newman

Mirabilis translates as "wondrous"—a very succinct, and apposite, one-word description of all of the many astonishing creatures investigated in this book.

At the fringes of traditional mainstream zoology is cryptozoology—the study of elusive controversial animals whose existence or taxonomic identity has yet to be formally recognised by science. Here, sea serpents and man-beasts, mystery big cats and desert-dwelling death worms, giant birds and pygmy elephants, surviving dinosaurs and lingering pterodactyls, monstrous snakes and colossal octopuses jostle for attention from the ever-expanding ranks of learned professional and well-informed amateur researchers pursuing them in remote and even sometimes not-so-remote regions of the world.

Since the late 1980s, well over a dozen books of mine on mystery beasts of many forms have been published, together with countless articles that have appeared in publications all over the world. Now, at last, here is a collection of some of my most significant, acclaimed articles on cryptozoological animals (cryptids), drawn from a wide range of magazines and periodicals as well as from my award-winning online blog, ShukerNature, but never previously collected together in book form. Moreover, whenever appropriate and possible, they have been fully revised and expanded to incorporate new, relevant information that has become available since their original publication; and they are illustrated with a wide range of images, many of which again have never before been published in book form (or, indeed, at all, in a number of cases).

Here you will make the acquaintance of such exotic and esoteric beasts as all manner of horrifically over-sized spiders, some the size of puppies or even bigger, that have been reported from many disparate corners of the globe, even including the U.S.; gigantic beavers that not only have emerged from lakes and rivers in many parts of North America but also may have stepped forth from prehistoric times into the present day; as well as an ocean of sea dragons and other extraordinary beasts once deemed to have existed in antediluvian times when fossils finally began to attract serious interest in pre-scientific times.

Also awaiting your attention within this book's prodigious ensemble of unnatural history is a vast array of fascinating but hitherto long-forgotten mystery beasts, such as the elephantine sukotyro of Indonesia, North America's baffling tygomelia that seemingly resembled an extraordinary crossbreed of giraffe and moose, a striped mystery horse from Spain, giant toothless sharks from South America, a missing micro-squirrel from Africa, and a very sizeable sea snail equipped with paws and antlers; plus a fascinating diversity of contentious cryptids from Madagascar that officially died out centuries ago but which, judging from detailed eyewitness testimony, may still exist on this immense "mini-continent" island, such as giant lemurs, dwarf hippopotamuses, and a number of other Malagasy exotica.

Nor should we (or could we!) overlook Trunko—the decidedly monstrous, hairy, elephant-trunked, and truly immense carcass washed ashore on a South African beach during the early 1920s and which perplexed scientists for generations until my recent discovery of its true identity; an extraordinary assortment of animals inexplicably discovered alive inside slabs of solid rock and other stony sepulchers; the bizarre appearance and entertaining evolution from medieval times to the modern day of horned hares and jackalopes; the extraordinary, still-unidentified giant chelonians and gargantuan crocodilians that have been reported from land, freshwater, and the seas all around the globe; the bemusing flying turtles and flower-engendered birds of China; the contentious skinned serpent of Christopher Columbus and the Hercynian unicorn encountered by Julius Caesar; a duplicitous pterodactyl dragon and Sonnerat's deceptive New Guinea penguins; the unique hippoturtleox of Tibet's Lake Duobuzhe; plus much, much more!

Let us not tarry even a moment longer. The miracles and marvels of *Mirabilis* await you impatiently inside, so I bid you welcome, and pray that your visit to my carnival of cryptozoology and unnatural history will be entertaining...and not *too* perilous!

Chapter 1:
THE MEGALOPEDUS, THE SUKOTYRO, AND A VERY CRYPTIC CABINET OF CURIOSITIES

> *Objects selected for a cabinet of curiosities sent down roots into the imagination, ancestral memory—part science, part myth. Potent objects—the large bones of a percheron, the huge horse on which a knight would ride; the crocodile, who... might just grow forever; volcanic ash from Mount Etna; or "metal vegetables," pieces of gold or silver that appeared on the surface of the earth—each of these objects would come into a collection as full of fact as of mystery.*
>
> — Rosamond Purcell, *Special Cases: Natural Anomalies and Historical Monsters*

It is always a delight to unearth accounts of hitherto-obscure mystery beasts from the archives of antiquity. Lately two such creatures, hailing from widely disparate geographical regions but of oddly similar appearance, have occupied my attention simultaneously, though with very different outcomes, as will be seen!

THE TUSKED MEGALOPEDUS

It was American cryptozoological contact Marc Gaglione who first brought the tusked megalopedus to my attention, asking me in early 2011 if I'd ever heard of this enigmatic beast:

> What are some animals, small or large, that are known only from a single specimen, skin, or skull? Is there something called a Tusked Megalopedus? I read the name somewhere.

However, I certainly hadn't, because I would have definitely remembered a name as distinctive as "tusked megalopedus," and, indeed, would have actively investigated it—if only because of the tantalizing cryptozoological coincidence that "megalopedus" translates as "bigfoot"! But a Bigfoot with tusks? Moreover, judging from the context of Marc's message, it appeared that this creature was known only from a single specimen, making it even more curious.

Mirroring that situation, internet searches for information concerning the tusked megalopedus yielded only a single but highly informative source. It stated that this mysterious creature had been described by no less an authority than Pliny the Elder (23-79 AD), the famous Roman naturalist-scholar and author of the encyclopaedic work *Naturalis Historia*. It also claimed that the only known specimen of the megalopedus was ensconced in a quite extraordinary cabinet of curiosities, and it included a description of this remarkable collection as witnessed by a visiting journalist that contains a brief account of that unique specimen:

> ...a strange stuffed animal greeted his eye: a large, tapirlike mammal with a huge muzzle, powerful forelegs, bulbous head, and curving tusks. It was like nothing he had ever seen before; a freak. He bent down to make out the dim label: *Only known specimen of the Tusked Megalopedus, described by Pliny, thought to be fantastical until this specimen was shot in the Belgian Congo by the English explorer Col. Sir Henry F. Moreton, in 1869...*[But] could it be true? A large mammal, completely unknown to science?

So where can this peerless repository of the world's only tusked megalopedus specimen, and untold other scientific treasures besides, be found? Yes, you've guessed it—in the pages of a novel! Published in 2002 and entitled (what else?) *The Cabinet of Curiosities*, it was written by Douglas Preston and Lincoln Child, and is the source of the above statements and quote.

Once I'd discovered that, I contacted Marc, who confirmed that this was indeed the book where he had read about the megalopedus, but he had wondered whether, although the story was fiction, the animal itself might have been genuine. Pliny documented many bizarre creatures in his encyclopaedia, some of which were later unmasked as legendary beasts or even wholly fictitious monsters with no basis either in fact or in folklore, yet others were indeed real, so such a situation would not be unprecedented.

As far as I am aware, however, there is no mention of the tusked megalopedus in any of Pliny's works. Another reason for doubting even an erstwhile existence for this ultra-cryptic creature supposedly hailing from the Belgian Congo is that the name of its discoverer, Col. Sir Henry F. Moreton (for whom no record of authenticity exists), is clearly inspired by that of a real explorer, Sir Henry Morton Stanley, who did indeed contribute to a major zoological revelation in the Belgian Congo—the okapi *Okapia johnstoni*, formally discovered in 1901.

Reconstruction of the tusked megalopedus (Tim Morris)

Nevertheless, I would be delighted to be proven wrong, and receive verifiable evidence that Pliny genuinely documented the mystifying megalopedus, or even something comparable upon which the megalopedus may have been based. Consequently, if anyone out there can do so, please send in details.

Having said that, there is one final, but extremely significant, twist in the tale (if not the tail!) of the tusked megalopedus. Even though this is assuredly a make-believe mammal, its description is strangely reminiscent of a seemingly genuine yet wholly obscure, long-forgotten mystery beast that I serendipitously uncovered during my megalopedus investigations. And the name of this overlooked oddity? The sukotyro of Java.

THE SUKOTYRO

While seeking possible images of the megalopedus before discovering its true nature, I spotted a very unusual antiquarian print for sale online. It depicted two animals. One was Australia's familiar duck-billed platypus. The other was an entirely unfamiliar hoofed mammal labeled as the sukotyro. This print was a color plate taken from Ebenezer Sibly's *A Universal System of Natural History* (1794). I collect antiquarian natural history prints as a hobby, but I was reluctant to purchase this one as I considered its asking price to be unjustifiably high. Happily, however, further internet perusal soon yielded several other plates depicting this same creature, all dating from the early 1800s.

Despite emanating from different sources, these plates' depictions were all clearly based upon some single, earlier, original illustration (see later). And as they were reasonably priced, I duly purchased no less than three large, excellent plates (two in color, one b/w) that included the instantly recognizable sukotyro image (together with various well-known beasts).

Remarkably, the sukotyro does not readily resemble any known mammal. Its large, burly body (variously portrayed as elephantine grey or deep-brown in coloured depictions) is somewhat rhinoceros-like in general shape but its smooth skin lacks these creatures' characteristic armour, and its long, bushy-tipped tail differs from their shorter versions. Equally distinct is the short upright narrow mane that runs down the entire length of its back.

Furthermore, each of its feet appears to possess four hooves (thereby allying it with the pigs, hippopotamuses, camels, ruminants, and other even-toed ungulates or artiodactyls, whereas the rhinoceroses are odd-toed ungulates or perissodactyls, which also include the horses and tapirs). Its head is also totally unlike that of any rhino, sporting a sturdy but elongate, hornless muzzle ending in a pair of decidedly porcine nostrils, a pair of long, pendant ears, and, most distinctive of all, a pair of truly extraordinary tusks.

Bizarrely, all of the sukotyro illustrations that I have seen depict these tusks emerging from a point located directly beneath and behind the eyes rather than from any anterior region of the jaws. In addition, they are held above the level of the snout, and (rather like those of elephants) are almost entirely horizontal in orientation with only their tips exhibiting a slight upward curve. I assume, therefore, that these depictions are meant to signify (albeit with poor artistry) that the sukotyro's tusks emerge from some site at the back of the (upper?) jaws.

Tusks aside, in overall appearance the sukotyro resembles a big wild pig, but where does (or did) it come from? What, if indeed anything, is known about this strange creature?

The earliest known description of the sukotyro is that of Johan Niewhoff (aka Niuhoff and Neuhoff) in his account of his travels to the East Indies, entitled *Die Gesantschaft der Ost-Indischen Gesellschaft in den Vereinigten Niederländern*, and published in 1669. He also included the illustration of it reproduced here, upon which all subsequent ones appear to have been directly based.

Niewhoff's description, kindly translated from Dutch into English for me by longstanding Dutch cryptozoological correspondent Gerard Van Leusden, reads as follows:

> The animal Sukotyro as it is called by the Chinese has a wonderful and strange shape. It is about as big as an ox, has a snout like a pig, two long rough ears and a long hairy tail and two eyes that stand high, completely different from those in other animals, alongside the head.
>
> At each side of the head, along the ears, are two long horns or tusks that are darker than the teeth of the elephant. The animal lives from vegetables and is seldom captured.

My continuing searches revealed that the sukotyro received its most authoritative scientific coverage in 1799, by British Museum zoologist George Shaw in Vol. 1 of his exhaustive 16-volume *General Zoology: Or Systematic Natural History*. Inserting it directly but somewhat hesitantly after the elephant in this volume's main text (and neglecting

The Megalopedus, the Sukotyro, and a Very Cryptic Cabinet of Curiosities

The sukotyro, as illustrated in Johan Niewhoff's 17th-century travelogue alongside an elephant and various other animals (public domain)

to mention, incidentally, that in 1792 its species had been formally christened *Sukotyro indicus* by fellow zoologist Robert Kerr), Shaw paraphrased Niewhoff's description and concisely documented this enigmatic mammal as follows:

> That we may not seem to neglect so remarkable an animal, though hitherto so very imperfectly known, we shall here introduce the Sukotyro. This, according to Niewhoff, its only describer, and who has figured it in his travels to the East Indies, is a quadruped of a very singular shape. Its size is that of a large ox: the snout like that of a hog: the ears long and rough; and the tail thick and bushy. The eyes are placed upright in the head, quite differently from those of other quadrupeds. On each side the head, next to the eyes, stand the horns, or rather teeth, not quite so thick as those of an Elephant. This animal feeds upon herbage, and is but seldom taken. It is a native of Java, and is called by the Chinese *Sukotyro*. This is all the description given by Niewhoff. The figure is repeated in Churchill's *Collection of Voyages*

and Travels, vol. 2. p. 360. Niewhoff was a Dutch traveller, who visited the East Indies about the middle of the last century, viz. about the year 1563 [*sic*—should be 1653], and continued his peregrinations for several years. It must be confessed that some of the figures introduced into his works are not remarkable for their accuracy.

This would presumably explain, therefore, the anatomically aberrant positioning of the sukotyro's tusks, and also, probably, the upright positioning of its eyes. Alternatively, if the latter are portrayed correctly, it could be suggested that the sukotyro spends time submerged in water, with only its eyes showing above the surface, as with hippopotamuses, whose eyes are also placed high on their skull. However, so too are the hippos' nostrils and ears, whereas those of the sukotyro are not, thereby reducing the likelihood that it does spend any length of time largely submerged.

As for the sukotyro's tusks: ignoring their potentially-inaccurate horizontal orientation, they remind me both in shape and in size of those bizarre versions sported by the Indonesian babirusas *Babyrousa* spp. (formerly treated as a single species, but recently split into four separate ones). These grotesque-looking wild pigs are famous for the huge vertical tusks sported by the males, in which not only the lower tusks but also the much larger upper ones project vertically upwards, with the upper ones growing directly through the top of the snout!

Babirusas are native to Celebes (Sulawesi) and various much smaller islands close by, but zoologists believe that they may have been deliberately introduced onto at least some of these latter isles by human activity rather than by natural migration. If so, might they also have been transported elsewhere in Indonesia, perhaps as far west as Java, in fact?

Following Shaw's cautious coverage of the sukotyro, other zoologists adopted an even more sceptical view of it. This deepened still further following the revelation that a pair of alleged sukotyro tusks acquired by British Museum founder Sir Hans Sloane during the 1700s were actually the horns of an Indian water buffalo. These had been presented as a gift to Sloane by a Mr. Doyle after he had discovered them in a partially worm-eaten state inside the cellar of a shop in Wapping, London, and were formally documented in 1727 within the *Memoirs of the Academy of Sciences*. Eventually, with no further specimens or data regarding it coming to light, the sukotyro was dismissed by the scientific world as a hoax, after which it quietly vanished from the natural history books. But was it really a hoax?

The more I look at the depictions of the sukotyro, the more they seem—at least to me—to resemble a distorted but still-identifiable portrait of a babirusa. There is no indication that any of these porcine species exist on Java today, but perhaps Niewhoff's mystifying sukotyro is evidence that one did exist there long ago. Alternatively, could this cryptid even have been an unknown relative of the babirusas, differing from them via its bulkier form and longer ears, but still recognizably akin?

19th-century engraving of a babirusa (public domain)

Finally, an even more controversial, dramatic possibility is that the sukotyro was a bona fide prehistoric survivor. If we assume that its extraordinary tusks really were horizontal, and not an artifact of poor artistry, its overall form calls to mind a modest-sized stegodont, a cousin of today's elephants, complete with proboscis (albeit rather short in the sukotyro images), floppy ears, and long hairy-tipped tail. Moreover, a dwarf, buffalo-sized species of stegodont, *Stegodon florensis*, is known to have survived on the island of Flores, just east of Java, until as recently as 12,000 years ago—so could a comparable form have existed on Java too, and lingered even longer there, right into historical times?

Sadly, we will probably never know, for in a very real sense the lost, forgotten sukotyro is today just as intangible as its literary, fictional doppelgänger, the tusked megalopedus.

Chapter 2:
MEGA-TURTLES AND TITANIC TORTOISES—
IN THE WAKE OF GIANT CRYPTO-CHELONIANS

A number of reports of giant sea turtles, apparently significantly larger than known species, do exist, however, and it is not impossible that a small population, breeding on an uninhabited group of islands, could have remained undetected.
— David Alderton, *Turtles and Tortoises of the World*

We first heard about the "dinosaur turtle" from Gilbert Bonguenele Manengue, the agent de la Sécurité Publique at Brazzaville...He reported that he had received his information from eyewitness observers living in the village of Boha; data obtained from him, therefore, were secondhand. Information about the great turtle was very limited, possibly because the only unusual feature about the animal was its colossal size. Informants may have felt it was quite enough simply to say "giant turtle" about an animal whose shell was some 4 to 5 metres (12 to 15 feet) in diameter.
— Roy P. Mackal, *A Living Dinosaur? In Search of Mokele-Mbembe*

Do the vast oceans of our planet conceal great sea turtles far larger than any that are officially known to exist there? Are there mega-giant tortoises living in reclusive sanctuary amid some of the more remote regions of the globe? Could freshwater turtles of exceptional size lurk beneath the water surface of lakes and rivers in secluded isolation from the prying eyes of science? Read the following selection of reports—the most comprehensive survey of giant crypto-chelonians ever assembled in print—and judge for yourself.

FATHER-OF-ALL-THE-TURTLES

The largest species of sea turtle ever known to have existed was *Archelon ischyros*, which lived some 70 million years ago during the Late Cretaceous Period in the seas around what is now North America. The largest specimen on record measured over 13 ft. long and roughly 16 ft. across from flipper to flipper. In comparison, the largest species known to exist today, the leathery (leatherback) turtle *Dermochelys coriacea*, attains a maximum recorded length of a "mere" 9.8 ft., but averages only 6-7 ft. However, reports of substantially larger sea turtles are also on file—veritable behemoths, in fact, which if their existence were ever scientifically confirmed would rival even the mighty *Archelon* itself.

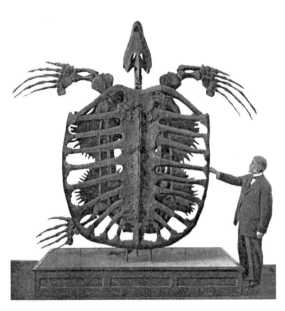

Skeleton of *Archelon* (public domain)

In his classic if highly controversial tome *In the Wake of the Sea-Serpents* (1968), Belgian zoologist Bernard Heuvelmans postulated the existence of a number of hypothetical creatures to explain hundreds of reports of sea serpents from around the globe. One of them was what he termed the "Father-of-All-the-Turtles"—a name originally given by native Sumatran fishermen to a traditional sea deity in which they vehemently believe, and which allegedly assumes the guise of a gargantuan marine turtle. Alternative sea serpent classification systems subsequently devised by other cryptozoologists have also included this category, and there are several notable eyewitness accounts on file indicating that a turtle of truly immense size may indeed have been encountered in various far-flung maritime localities.

GIANT TURTLES IN ANTIQUITY AND MEDIEVAL TIMES

Perhaps the earliest report dates back as far as the 3rd century AD. This was when Roman scholar Claudius Aelianus (popularly known merely as Aelian), writing in his 17-volume treatise *De Natura Animalium*, referred to the existence in the Indian Ocean of turtles so colossal in size that their huge shells—said to be as much as 23.5 ft. in circumference—were sometimes used by the native people as roofing material! Modern-day sceptics have claimed that if such shells truly existed, they must have been fossils. And it is certainly true that portions of fossilised shells from the prehistoric giant tortoise *Colossochelys atlas* have been unearthed in the rich deposits of Nepal's Siwalik Hills. Yet fossil shells would surely have been too brittle and much too heavy for roofing purposes.

Writing in his own magnum opus, *Geography*, which he completed in 1154 AD, Muhammad al-Idrisi, a notable Moroccan Islamic traveller, cartographer, and archaeologist, referred to comparably immense turtles, up to 20 cubits (33 ft.) long, living in the Sea of

Herkend, off the west coast of Sri Lanka, whose females contained up to a thousand eggs. Although he never personally visited Asia, he collated considerable amounts of detailed information from Islamic explorers and merchants, and recorded on Islamic maps. Having said that, however, turtles generally lay no more than a hundred eggs at a time, not a thousand, so perhaps some such reports were exaggerated. (Could this also account for the huge size claimed for these Herkend turtles?)

CHRISTOPHER COLUMBUS AND ANNIE L. HALL

Possibly the most famous eyewitness of a reputed mega-chelonian was none other than New World discoverer Christopher Columbus. In early September 1494, along with several of his crew, he witnessed an extraordinary creature likened to a whale-sized turtle with a visible pair of flippers and a long tail that kept its head above the water surface while swimming by as his vessels were sailing east along the southern coast of what is now the Dominican Republic.

On March 30, 1883, while aboard the schooner *Annie L. Hall* in the North Atlantic's Grand Banks, Captain Augustus G. Hall and his crew spied what they initially took to be an upturned vessel but which, when they approached to within 25 ft. of it, proved to be an enormous turtle. By comparing its dimensions with those of their vessel, they were able to estimate its total length as being at least 40 ft., its width as 29.5 ft., and its height from its carapace's apex to its plastron or under-shell's most ventral point as 29.5 ft. Even its flippers were immense—each one approximately 20 ft. long. Not surprisingly, the captain deemed it inadvisable to attempt capturing this shell-bearing leviathan!

TURTLES OF A (VERY) DIFFERENT COLOUR!

During the 1950s, two ultra-giant turtles were reported that were highly distinctive due not only to their great size but also to their very unusual colouration. One of these was an alleged 14-ft.-long yellow turtle witnessed on March 8, 1955, by L. Alejandro Velasco while stranded on a raft off Colombia's Gulf of Urabá. Its claimed length is greater than that of the largest leathery turtle on record. And the other was a 44-ft.-long pure-white turtle with 14-ft. flippers that was sighted south of Nova Scotia, Canada, in June 1956 by crew on board the cargo steamer *Rhapsody*. According to their account, this huge creature could raise its head 8 ft. out of the water.

ENCOUNTERING THE SOAY BEAST

Perhaps the most famous and controversial modern-day sighting of an alleged giant sea turtle occurred just a few years later, off the Scottish Inner Hebridean island of Soay. On September 13, 1959, while fishing here for mackerel, holidaying engineer James Gavin and local fisherman Tex Geddes were very startled to observe an extremely large sea creature swimming directly towards their boat, its head and back readily visible above the sea surface, until it was no more than 60 ft. away.

According to their description, documented in a major *Illustrated London News* report of June 4, 1960, the head of this remarkable beast was definitely reptilian and resembled a tortoise's, with lateral eyes and a rounded face plus a horizontal gash for a mouth when closed, but it was as big as a donkey's, and the neck was cylindrical. The exposed portion of its back was humped in shape, and running down the centre was a series of triangular-shaped spines or serrations, like the teeth of a saw. The animal was so close that when it opened its mouth to breathe, emitting a very loud whistling roar, they could see the red lining inside, and what looked like loose flaps of skin hanging down from the roof, but there were no teeth.

After remaining in sight for five minutes, their extraordinary visitor dived forward and vanished beneath the water surface, then emerged again almost a quarter of a mile away, before they watched it disappear completely. Moreover, the crews of two lobster boats, fishing north of Mallaig on the Scottish mainland, also spied this mysterious creature, much to their alarm!

Due to its mid-dorsal serrations, British zoologist Maurice Burton wondered whether it may have been an escapee iguana, but the remainder of the Soay beast's description does not correspond with such an identity at all, and much more readily recalls a chelonian. Moreover, certain terrapins even possess dorsal serrations, though terrapins are, of course, freshwater species, not marine.

COULD UNDISCOVERED MEGA-TURTLES EXIST TODAY?

Reading through these reports, various objections to the possibility of giant turtles existing undiscovered by science soon come to mind, but are they insurmountable? For instance: all known species of sea turtle have only short tails, so the long tail of the specimen sighted by Columbus and his crew is unexpected. Yet there are no sound anatomical or physiological reasons why a long-tailed species might not exist. In any case, it may be that at least part of the tail's length was an optical illusion, caused by the wake created as it swam by.

The odd colours of the two specimens reported during the mid-1950s pose another problem. But perhaps the yellow colouration of the Colombian individual was merely due to reflected sunlight, or it might have been a rare xanthic (yellow mutant) individual. Equally, the white *Rhapsody*-spied turtle may conceivably have been an albino or leucistic specimen—such specimens have been reported from many reptilian species, even ones as large as crocodiles and alligators. Indeed, a spectacular photo of a leucistic specimen of black (aka green) sea turtle *Chelonia mydas* snapped by wildlife photographer Jeff Lemm in 2012 was lately posted on a herpetological website (fieldherpforum.com).

The cold-water regions in which the *Rhapsody* specimen and the Soay beast were observed argues on first sight against their being reptiles. However, the leathery turtle is famed for its ability to withstand cold-water temperatures, and of particular note in relation to the Soay case is the remarkable but fully confirmed fact that in August 1971 a leathery turtle was caught near Mallaig!

If mega-turtles do truly exist, they must be at least predominantly pelagic in occurrence, otherwise they would have been seen far more frequently. Yet even if this is so, surely they would have been observed on land at some point, coming ashore to lay their eggs? Possibly—then again, there are countless small, remote, uninhabited tropical islands and island groups that have never been visited by humans. Indonesia, for instance, consists of more than 17,500 islands, and more than 7,100 constitute the Philippines.

So perhaps it is beneath the coastal sands of tiny isles such as these, emerging at night for just a few hours only once every 2-3 years and far removed from prying eyes or from other potential sources of danger to their precious offspring, where these shy reptilian giants (so adept when at sea but so vulnerable when on land) entrust their precious eggs and, while depositing them, their own lives too, before returning once more to the safety of their vast maritime domain.

Who knows? One day, possibly, an intrepid adventurer may visit one of these anonymous specks of land and there encounter a trail of huge flipper prints left behind by some great chelonian Man Friday.

Even so, the seas do not have a monopoly on extra-large cryptic chelonians. Some equally fascinating examples have also been reported from a diverse range of terrestrial localities.

GONE IN THE GALAPAGOS, OR JUST OVERLOOKED?

It might seem difficult to lose anything as large as a giant tortoise, but this ostensibly implausible feat may have happened on more than one occasion, and to more than one subspecies, in that most famous of tortoise-associated archipelagos, the Galapagos.

Owned by Ecuador and situated just over 600 miles to the west of this South American country's Pacific coast, the Galapagos Islands take their name from their most famous, and sizeable, native inhabitants—the giant tortoises. At one time, many different species were recognized, but nowadays these are all deemed to be subspecies of a single species, *Chelonoidis nigra*. Fifteen subspecies are generally recognized, of which at least six are nowadays deemed wholly extinct or extinct in the wild—but are they?

Until his death from unknown causes in June 2012 at the age of approximately 100, the most famous living Galapagos giant tortoise was Lonesome George, the last known surviving specimen of the Pinta (=Abingdon) Island subspecies, *C. n. abingdoni*. Noted for its distinctive saddle-shaped shell, very different from the more typical domed carapace of most tortoises, the Pinta Island giant tortoise was once common but scientists assumed that it had been hunted into extinction for its tasty meat by the 1930s, because expeditions to this island during that decade and again during the 1950s failed to find a single specimen.

In 1964, however, a new expedition to Pinta was startled to encounter no less than 28 dead tortoises there, all of which appeared to have died no earlier than five years previously. This remarkable discovery meant that they must all have been alive, but very effectively concealed, during those expeditions of the 1930s and 1950s. Bearing in mind that Galapagos giant tortoises can exceed 6 ft. in length and 880 lb in weight, for almost 30 of these slow-moving reptilian goliaths to avoid detection by several teams of researchers specifically looking for such animals is quite an achievement—but an even greater surprise was still to come.

Mega-Turtles and Titanic Tortoises

19th-century engraving depicting Pinta Island giant tortoises (public domain)

In March 1972, a single *living* specimen was found on Pinta. An adult male, he was swiftly taken for safekeeping to the Charles Darwin Research Station on Santa Cruz, a much larger Galapagos island, where he lived on as a chelonian celebrity for the next 40 years, dying during the early morning of June 24, 2012. Despite further searches, however, no additional specimens were discovered on Pinta, and so this lone individual was eventually dubbed Lonesome George. But was he really the last Pinta Island giant tortoise? In 1981, tortoise droppings just a few years old were unexpectedly found here—and if there are droppings, there must have been at least one tortoise on the island to produce them.

Moreover, in 2006 it was announced that a giant tortoise called Tony living at Prague Zoo in the Czech Republic might be a purebred Pinta Island giant tortoise! Extensive genetic analyses are still underway to determine his precise taxonomic identity. And on Isabela (=Albermarle) Island, a first generation (F1) male hybrid specimen of Pinta Island giant tortoise was found in 2007. This in turn means that somewhere on Isabela, there may be a surviving, undiscovered, purebred *C. n. abingdoni*. If so, it must have been introduced here from Pinta, this subspecies' only native locality. If resurrecting Lonesome George's subspecies as a viable one is not reward enough in itself, the Charles Darwin Research Station is offering $10,000 to anyone who can locate a female purebred Pinta Island giant tortoise.

As for Lonesome George, it is presently unclear whether his tissues will be preserved for possible cloning attempts in the future. Although he had been loyally cared for by a personal keeper throughout his four decades at the institute, it was with ironic inevitability that he was alone when death finally released him from so many years of isolation from all other members of his kind.

More mysterious even than Pinta's giant tortoise is that of Fernandina (=Narborough) Island—*C. n. phantastica*. This enigmatic subspecies is known only from a single preserved specimen, spied (and killed!) there in 1906 by Rollo H. Beck while leading an expedition from the California Academy of Sciences. Due to this specimen's unique status, many herpetologists no longer deem *C. n. phantastica* to be a valid subspecies, arguing that others would surely have been discovered by now, and that it must have been introduced onto Fernandina Island from elsewhere. But, in 1964, several fresh fecal droppings from some unseen tortoise(s) were found on this island's southern slopes, indicating that Beck's specimen may not be unique after all.

Also officially extinct is the Floreana (=Charles) Island giant tortoise *C. n. nigra*. As recently as January 2012, however, more than 80 hybrids of this "lost" subspecies and the Isabela Island giant tortoise *C. n. becki* were unexpectedly discovered on Isabela. As a result, researchers now consider it plausible that around 38 purebred Floreana Island giant tortoises are present incognito on Isabela, probably transported here from their native Floreana by whalers.

BORN AGAIN IN THE SEYCHELLES—AND MADAGASCAR?

The Galapagos Islands are not the only locality in modern times to be inhabited by giant tortoises. Almost as famous for this same zoological reason is the Indian Ocean island of Aldabra. A coral atoll in the Aldabra Group, it is home to *Aldabrachelys gigantea* (aka *Dipsochelys dussumieri*), a mega-tortoise that can be as large as the Galapagos giant tortoise.

The Aladabra Group is part of the Seychelles archipelago, and until little more than a decade ago, *A. gigantea* was the only species of giant tortoise thought to survive anywhere there. Additional Seychelles species had been described, but these were all thought to have died out by the mid-19th century.

But in the opinion of many researchers, detailed morphological and genetic studies from the late 1990s of Aladabran giant tortoises in captivity show that some of these specimens actually belong to one or other of two "lost" Seychelles giant tortoise species—the hololissa *A. (=D.) hololissa* and Arnold's giant tortoise *A. (=D.) arnoldi*. So where there was once only one surviving Seychelles species, there are now three, and the two resurrected ones are currently the subjects of a captive breeding and reintroduction program run by the Nature Protection Trust of Seychelles.

Another Indian Ocean locality that was once frequented by giant tortoises is Madagascar. Its principal species was formally named (as *Testudo grandidieri*) and described in 1885,; it was well over 3 ft. long, more than a ton in weight, and is nowadays referred to as *A. (or D.) grandidieri*. Thought to have enjoyed basking in the sun, half-submerged in shallow water, it is currently known only from remains (bones, shells, etc) dating 2290-1250 BP (Before Present), but may have survived into much later times. Indeed, as recently as 1950, Raymond Decary included within his authoritative work *La Faune Malgache* the following brief but very tantalizing snippet of information:

> It has been established that a mysterious animal which may be a large tortoise lives in certain lakes in the caves in the south-west; is it a last survivor of *Testudo grandidieri*?

Sadly, more than half a century later, his very thought-provoking question is still unanswered.

A PUZZLE FROM THE PARACELS

Located in the South China Sea, the Paracels constitute a group of islands whose ownership has inspired heated dispute between China, Taiwan, and Vietnam down through the years, but they are currently administered by China's Hainan Province. On January 30, 2013, cryptozoological researcher Richard Muirhead brought to my attention a very intriguing report that had appeared in the *China Mail* newspaper on June 18, 1947.

This report stated that a 1000-year-old giant tortoise had just died at Canton Zoo and was originally from the Paracel Islands. Needless to say, the alleged age for this specimen was evidently a very sizeable exaggeration, but to my mind what was far more interesting anyway was the tortoise's provenance—because no species of giant tortoise has ever been confirmed from the Paracels.

Far more likely, therefore, is the possibility that the Canton Zoo specimen had been brought there from elsewhere by Chinese seafarers, and its most plausible origin was the Seychelles archipelago. However, as Richard also revealed:

> [In] J. Pinkerton *A General Collection of...Voyages and Travels* (1814) a group of four shipwrecked sailors survived on a great quantity of tortoises between 1690 and 1701. A pro-government Vietnamese web site (in English) cites a historian Le Qui Don (1726-1784) who mentioned [extremely] large tortoises there.

So might there once have been giant tortoises on these inconspicuous islands, but which, just like their Galapagos compatriots, were slaughtered for their tasty meat by sailors? With no specimen to examine, we may never know—unless of course some physical remains of the Canton Zoo specimen were preserved...?

THE GIANT STONE TORTOISES—A MONGOLIAN MYSTERY

In the Russian republic of Buryatia (in southcentral Siberia), the Russian Far East, the Korean Peninsula, and, particularly, Mongolia, a number of very large stone statues of tortoises can be found—but these are no ordinary tortoises, as documented by chelonian expert and palaeobiologist Viacheslav Mikhailovich Chkhikvadze, from the Georgian Academy of Sciences in Tbilisi.

In a paper from 1988, Chkhikvadze focused much of his attention upon three such statues in Mongolia, still standing amid the palace ruins at Karakorum, which was once the Mongol Empire's capital city. The biggest of these statues dates back to the 13th century, and measures 92 ft. long, 34.5 ft. wide, and 41 ft. high. The Mongolian name for it sounds like "jastmel'chij." They all portray a type of huge long-shelled tortoise with a giant head on a thick neck, protruding eyes, and—incongruously for a tortoise—a pair of short, closely-pressed ears. Equally odd are the sharp teeth that project from the upper jaw (chelonians are normally toothless), and all four feet have five long claws. One of the statues incorporates an inset seat for visitors to sit upon.

A particularly old example of these statues, dating from the 8th century, was found in Mongolia's northern province, and has subsequently become the subject of much study at the archaeological department of the Academic History Institute in Ulan Bator. Moreover, there were further examples at each gate on the four sides of the city wall at Karakorum. Steles (pillars) were present on the backs of these tortoises, which were crowned with beacons for travelers in the steppe.

Perhaps the most famous of the giant stone tortoises outside Mongolia is the example discovered by early Russian settlers in 1868, on the grave mound of the Jurchen general Asikui, where it had been installed in c.1193. Originally sited near the village of Nikolskoye (today the city of Ussuriysk) in what is now the Russian Far East, in 1895 it was transported to the I.N. Grodekov Khabarovsk Territorial Museum of Regional Studies. It dates from the Jin Dynasty (1115-1234).

What made them so intriguing to Prof. Chkhikvadze is that in his view, their basic tortoise form that is revealed once its stylised non-chelonian features—ears, teeth—are stripped away bears a remarkable similarity to the American snapping turtles. In prehistoric times their taxonomic family, Chelydridae, was distributed widely in Europe, Asia, and the New World, but today it is confined exclusively to the Americas (except of course for snapping turtle escapees/releases from captivity, several of which have been found living in British and other European ponds and lakes during recent years).

Consequently, as noted by Chkhikvadze, it remains a riddle how Central Asiatic sculptors in the Middle Ages knew a type of tortoise that in the Old World had been considered extinct for the past two million years. There is a remote possibility that these sculptors had seen fossilized remains of prehistoric Asian chelydrids. Unusual or eye-catching fossils have inspired a number of medieval representations in a wide range of localities globally, so such a situation would hardly be unique. Also remote, but again not impossible, is that a species of large Asian chelydrid survived into much more recent times than is currently known from the fossil record, and either directly or via generations of verbally-transmitted descriptions inspired the sculptors responsible for the statues.

Interestingly, the carving of giant stone tortoises in the Far East underwent an evolution of its own down through the ages, witnessing as already noted the popular addition of a stele to the tortoise's back, so that it became a symbol of heavy loads and burden. And in more modern times, especially in China where dragons have always played such a significant and diverse role in this country's mythology, the giant tortoise has itself acquired the head of a dragon—yielding a ferocious-looking hybrid monster known as a bixi.

AMPHICHELYDIDS AND AN ANOMALOUS PORTRAYAL IN PERU

Such explanations as these may also account for another anomalous portrayal, this time at Marcahausi, which is a plateau 13,000 ft. above sea level, and situated some 50 miles northeast of Lima, Peru. It was here that in 1952 Dr. Daniel Ruzo discovered a series of fascinating megalithic sculptures. Not only did they include some enormous human figures carved from the rocks but also a wide range of animals, including lions, camels, elephants—and one that in Ruzo's opinion seemed to represent an amphichelydida—suborder of prehistoric chelonians characterized by their non-retractable neck.

Clearly, the discovery of unexpected forms of giant tortoise in the future would not be unprecedented. So we must hope that searches for such creatures are indeed conducted. One thing is certain: unlike so many mystery beasts, it is highly unlikely that these cryptic reptiles will be speedily running away from their seekers!

Having now reviewed a diverse selection from the seas and on land, it is time to investigate some over-sized but under-explained chelonians from a variety of freshwater localities around the world.

THE CONGOLESE NDENDECKI—A DINOSAUR TURTLE?

During the 1980s, Roy P. Mackal, a now-retired University of Chicago biochemist and enthusiastic spare-time cryptozoologist, led two expeditions to the People's Republic of the Congo (formerly the French Congo) in search of a mysterious water beast known as the mokele-mbembe, which apparently bears a remarkable resemblance to a small sauropod dinosaur. Sadly, they did not encounter any such creature while there, but they did discover that it was only one of several different types of very large mystery reptile claimed by the local pygmies and Western missionaries to inhabit the vast and virtually inaccessible Likouala swamplands in this country's central region.

Among this assortment of cryptozoological curiosities was the ndendecki, which Mackal's expeditions nicknamed the dinosaur turtle due to its huge size, at least according to reports from eyewitnesses living in the village of Boha and collated by Gilbert Bonguenele Manengue. He was a public security agent commissioned by the military governor general of Brazzaville (this country's capital) to gather together for Mackal's expeditions any available information appertaining to putative unknown animals, and was himself born in Boha.

The shell of this gargantuan freshwater chelonian was claimed by

Reconstruction of the ndendecki (David Miller/Roy P. Mackal)

the Boha villagers to be conspicuously rounded in shape and to measure some 12-15 ft. in diameter, which is very considerably bigger than that of any known species of aquatic chelonian alive today. Happily, it apparently poses no threat to humans, browsing harmlessly upon detritus at the bottom of rivers and lakes.

As it happens, the Likouala swamps are already known to harbor one fairly large species of freshwater turtle, the African softshell *Trionyx triunguis*—a well-documented chelonian found in many parts of Africa. However, this species rarely exceeds 3 ft. across. Nevertheless, Congolese zoologist Marcellin Agnagna, who has participated in several mokelembembe expeditions and has extensive knowledge of the Likouala's fauna, is convinced that this species and the gigantic ndendecki are one and the same, dismissing the appreciable size difference between the two as nothing more than exaggeration on the part of the eyewitnesses.

Yet such exaggeration would need to be on a very grand scale indeed to enlarge a 3-ft. turtle by a factor of approximately four to five, and it seems unlikely that locals would be prone to such an overestimation of size for a species that is very common here and must therefore be a familiar sight to them. Consequently, it seems more plausible to believe that amid the Likouala's immense and yet scarcely explored wilderness of lakes, rivers, and swamps are some extra-large individuals of *T. triunguis* that have lived to great ages and continued growing throughout their protracted lives. Perhaps 13-16.5 ft. is indeed an exaggeration, but even a softshell only half as big would still be an extremely imposing creature to encounter, and could certainly explain the ndendecki.

THE BEAST OF BUSCO—A SNAPPY SECRET

Fulks Lake is situated near Churubusco (often shortened to Busco) in Indiana, and for over a century reports of an enormous turtle inhabiting it have surfaced—as, indeed, has the mystery reptile itself from time to time.

It all began in 1898, when Oscar Fulk claimed to have spied such a creature in this lake, followed by other eyewitnesses making similar allegations in 1914 and again in the 1940s. By then, the Busco beast had become such a focus of media attention that when an animal fitting this description was supposedly sighted in July 1948 on a nearby farm owned by Gail Harris, a multi-pronged search for it was eventually launched by Harris in March 1949. This lasted many months, employing scuba divers to look for the creature underwater, a dredging crane to capture it if spotted, a sump pump to drain the lake, and even a female sea turtle to lure it.

At first, things went well, with a wooden stockade being set up that trapped a monstrous turtle in about 20 ft. of water. Unfortunately, it then broke out of this enclosure. During the continuing search, numerous people visited Harris's farm hoping to spy the Busco beast, and on October 13, 1949, many observers saw a turtle as big as a dining room table leap from the water while trying to seize a duck being used as bait. Even so, it still evaded all attempts to capture it, and when Harris fell ill with appendicitis in December 1949 the hunt was abandoned.

Less than a year later, a truly immense freshwater turtle of undetermined zoological identity (but with a head said by witnesses to be as large as a human's) was spotted swimming around a drain leading into the Little Calumet River after a swamp near Black Oak, Indiana, was emptied. What could these mega-turtles have been?

The most popular theory is that they were extremely large specimens of the alligator snapping turtle *Macroclemys temminckii*. The largest living species of freshwater chelonian in North America, it is known to attain lengths of up to 2.5 ft. and weights of as much as 250 lb, possibly even more. Characterized by its huge head, as well as by its prominent dorsal ridges and hooked jaws, this species is indigenous to much of the southern United States, and spends most of its time in the water, but adult nesting females sometimes venture onto open land. As with the ndendecki and *T. triunguis*, an over-sized specimen or two of the alligator snapping turtle could adequately explain the elusive Busco beast.

TIBET'S HIPPOTURTLEOX—A VERY CURIOUS COMPOSITE

Due to the political situation, much of Tibet's terrain and freshwater lakes are little-known or off-limits to Westerners. Occasionally, however, a highly intriguing report emerges that suggests there may be some fascinating zoological finds waiting to be made here.

One such report briefly hit the headlines worldwide in September 1984, a mere 12 years after the creature itself had been captured. It claimed that back in 1972, a truly bizarre beast had been caught alive in Tibet's Lake Duobuzhe. It was described as having an ox-like body with

Reconstruction of the hippoturtleox (Tim Morris)

hippopotamus-like skin, the legs of a turtle, and a pair of short curled horns on its head. As a result of its composite morphology, it was duly dubbed a hippoturtleox by American cryptozoologist J. Richard Greenwell when he documented it in the Spring 1986 issue of the International Society of Cryptozoology's newsletter.

Tragically, however, following its capture this extraordinary-sounding creature was shot and bayoneted to death by some Chinese soldiers, then dragged to a nearby village, but the subsequent fate of its scientifically-priceless carcass is unknown (though it is quite likely to have been eaten—an ignominious fate shared by several other cryptozoological specimens down through the years). Nothing more has ever been reported about Lake Duobuzhe's hippoturtleox, and nothing like it has ever been documented since. In addition, I have been unable to locate Lake Duobuzhe online or in any atlas either, although there are literally thousands of lakes occurring on the Tibetan plateau, so this is not necessarily surprising. It is believed that the area used to be under the ocean, which then retreated, explaining the presence of so many lakes there today, a number of which are filled with saltwater.

Assuming the story was true, what could this beast have been? Its body's general shape, the appearance of its skin, and certainly its limbs are all reminiscent of a large freshwater chelonian—but if this is indeed what it was, how can its horns be explained? Could they have been a pair of specialized respiratory snorkels, perhaps? Or might they even have been real horns? After all, a horned chelonian would not be without precedent.

Until as recently as 2,000 years ago, the island of New Caledonia, situated to the east of Australia, was still home to an extraordinary cryptodire tortoise called *Meiolania mackayi*, whose skull bore a cluster of protuberances, including a pair of large, laterally-pointing horns. Much bigger *Meiolania* species formerly lived in Australia too (*M. brevicollis*), as well as on Lord Howe Island (*M. platyceps*), and Vanuatu (*M. damelipi*), but these all became extinct before their New Caledonian counterpart.

Without a body, preserved tissues, a photograph, or even an eyewitness drawing to examine, Tibet's hippoturtleox seems destined to remain an unclassifiable anomaly within the chronicles of cryptozoology—unless, one day, a second specimen appears. If it does, we can but hope that it will be treated more humanely than its mystifying species' previous representative.

THE GIANT TURTLES OF HANOI—LEGENDS COME TO LIFE

At the centre of Hanoi, Vietnam's teeming capital city, is a small lake called Hoan Kiem, which for at least five centuries has been rumored to house giant turtles. Yet even though several alleged sightings have been reported here in modern times, scientists have always discounted these creatures as mere legends, until March 24, 1998, that is, when a passing cameraman successfully filmed three such turtles surfacing to gulp air. After studying the film, Hanoi National University biologist Ha Dinh Duc announced that he considered them to represent a hitherto-undescribed species, which in 2000 he formally christened *Rafetus leloii*. Some scientists, conversely, challenged this classification.

Peter P.H. Pritchard, co-chairman of the Tortoise and Freshwater Turtle Specialist Group, visited the lake to view them personally. Afterwards, he revealed that although he did not dismiss out of hand the possibility that they were a new species, they may well constitute an outlying population of the exceedingly rare Chinese softshell *R. swinhoei* (currently known only from three living specimens). In his view, however, they were definitely not conspecific with the New Guinea giant softshell *Pelochelys bibroni*—an identity favored by several other herpetologists (even though this species' existence is currently confirmed only in New Guinea).

Pritchard's conclusions were echoed by Patrick P. McCord, a leading American expert on softshell turtles. Moreover, in 2003, herpetologists B. Farkas and R.G. Webb published a paper in which they denounced *R. leloii* as an invalid species, reclassifying the Hoan Kiem trio as representatives of *R. swinhoei*.

Regardless of their taxonomic identity, but incorporating data obtained from a stuffed Hoan Kiem specimen preserved in 1967 (after having been killed by a fisherman) and currently on display in Ngoc Son Temple on Jade Island near the lake's northern shore, Duc considers this lake's turtles to be the world's largest living freshwater chelonians. He cites for them a total length of approximately 6.5 ft., a width of 3 ft., and a weight of around 440 lb. Having said that, some researchers deem the narrow-headed softshell *Chitra indica* and the Asian giant softshell *P. cantorii* to be slightly bigger. The Hoan Kiem turtles' oval-shaped carapace is greenish-brown in color and mottled, their undersurface is pink, their football-sized head greenish-yellow, and their mouth downturned in form.

Whichever species is biggest, no one can deny that the existence of such a sizeable form of freshwater chelonian in the very heart of urban Hanoi that remained overlooked by science until as recently as the late 1990s is truly remarkable. Certainly, its official discovery and representation by at least one living specimen (the other two have not been conclusively sighted for several years now), which proved to be a female when it was rescued and cleaned in 2011, plus the preserved temple-ensconced individual, is among the most surprising and interesting herpetological revelations of modern times.

After all, it's not every day that a centuries-old legend comes to life.

Chapter 3:
HORNED HARES AND JACKALOPES —AN (UN)NATURAL HISTORY

A sporting goods dealer in a small Canadian town almost stopped the traffic recently with an unusual exhibit in his shop window. This piece of tomfoolery was a horned cotton-tailed rabbit. Some taxidermist had inserted the frontal bone and spiked horns of a gazelle into the dead coney's pate. Horned rabbits are by no means confined to the Americas, for in one of those fascinating Seine-side, open-air bookstalls in Paris I discovered, to my intense delight, a series of old German prints beautifully painted by an artist called Haid and dated 1794. This series included pictures of a number of antlered hares. The antlers of these non-existent creatures appeared to resemble those of immature roebucks.
— Henry Tegner, *The Countryman*, vol. 80 (Summer 1975)

Have you ever seen a hare with horns, or a rabbit with antlers? No? Neither have I—or, at least, I've never seen a living one, but I've seen quite a few preserved, taxiderm specimens. Are they real, though? Well, sort of.

A HISTORY OF THE HORNED HARE

As recently as the late 18th century, the authors of many of the early pre-scientific animal encyclopedias, or bestiaries as they were called then, still believed in the existence of fabulous beasts that nowadays have long been dismissed as non-existent fauna of folklore and legend—such as unicorns, dragons, satyrs, and mermaids. Another of these now-discounted creatures, far less dramatic than those listed above, yet no less intriguing, and often depicted in bestiaries, was the horned hare.

Despite its name, however, illustrations of this remarkable animal usually portrayed it as being much more rabbit-like than hare-like, and its "horns" were in fact antlers, branched at their tips, and frequently very similar in overall appearance to those of young roe deer. This bizarre mini-beast was widely reported across Europe but was said to be particularly abundant amid the forests of Bavaria in Germany. Indeed, its reality was so readily accepted by naturalists

Horned hare engraving from 16th century (public domain)

at that time that it even received its own formal Latin name—*Lepus cornutus* ("horned hare"). A number of highly prized stuffed specimens also existed, usually proudly displayed in hunting lodges or in private collections of unusual natural history exhibits known as cabinets of curiosities.

By the 19th century, conversely, advances in zoological research and scientific knowledge had shown that the horned hare was not only nonsense but a fraudulent one. Close examination of the taxidermy specimens revealed that they were hoaxes created by the skilful manipulation of large stuffed rabbits or hares onto whose heads had been craftily grafted pairs of young, short deer antlers, or, more rarely, the trimmed, pointed horns of small African antelopes, particularly duikers.

Two wolpertingers (Markus Bühler)

WOLPERTINGERS, RABBIT-BIRDS, AND SKVADERS

Yet even though its subject's authenticity had been disproved, the cult of the horned hare remained very much alive, so much so that by the mid-1800s a new craze had begun in earnest—the deliberate creation of ever more fanciful and exotic-looking horned hares, sporting not only antlers but even feathered wings, plus huge fangs startlingly similar to those of the prehistoric saber-tooth tigers! In Bavaria, such incredible composites were referred to as wolpertingers, and were often created specifically as tourist attractions, or as souvenirs to tempt and fool wealthy but gullible visitors. Even today, they appear on t-shirts and postcards, and privately owned specimens occasionally come up for sale in specialist auctions, where they invariably sell for very appreciable sums. The German Hunting and Fishing Museum at Munich houses a permanent exhibition of wolpertingers.

Equally bizarre is the rabbit-bird. Uniting the furry head of a rabbit with the feathered body of a bird, this highly unlikely hybrid was nonetheless firmly believed to be genuine by none other than the celebrated Roman naturalist-scholar Pliny the Elder (23-79 AD), who even documented it in his massive treatise *Naturalis Historia,* in which he claimed that it inhabited the lofty mountain peaks of the Alps. Needless to say, no such beast did, or does, exist—until 1918, that is.

For that was the year when Swedish taxidermist Rudolf Granberg created a stuffed hare with wings known as a skvader, deftly combining the head, foreparts, and limbs of a hare with the back, wings, and tail of a female capercaillie *Tetrao urogallus*, a very large species of grouse. Still exhibited today at the museum at Norra Berget in Sundsvall, eastern Sweden, it was inspired by an infamously far-fetched claim by hunter Håkan Dahlmark that he had shot just such a beast back in 1874 while hunting north of Sundsvall. Since then, other taxiderm skvaders have been created and, just like their Bavarian wolpertinger brethren, remain popular today.

AL-MI'RAJ—THE MALEVOLENT UNICORN HARE

Nor should we forget the astonishing al-mi'raj or unicorn hare. According to a number of medieval bestiaries, this enigmatic creature resembled a large yellow hare, but its brow bore a single unicorn-like horn, black in color. Said to inhabit a mysterious Indian Ocean island, and often featuring in Islamic poetry, this deceptive beast behaved in so placid and tame a manner that many a curious onlookers would approach it for a closer look—whereupon the al-mi'raj would suddenly lower its head and charge directly at its unsuspecting observer, fatally impaling him with its horn, then devouring him entirely!

A typical jackalope (CFZ)

JACKALOPES A-PLENTY!

But if you thought that rabbits with antlers or hares with horns were unique to Eurasia, think again! In North America, they have their own, exceedingly famous (albeit equally fraudulent) counterpart—the jackalope. There, hares are referred to somewhat confusingly as jack rabbits, and this continent is also home to a unique hoofed mammal called the pronghorn or pronghorn antelope *Antilocapra americana* (although, taxonomically speaking, it is not a true antelope but instead the only surviving species in an entirely separate ungulate family that in prehistoric times yielded a wide diversity of forms). Countless restaurants, bars, and local museums all over the U.S. and Canada boast stuffed specimens or shield-mounted trophy heads of jackalopes, created by affixing the horns of pronghorns, or sometimes the antlers of young deer, to the heads of stuffed jack rabbits.

Jackalope folklore dates back centuries, to when the first lumberjacks sat by their outdoor log-fires in their campsites at night, and told tall tales and yarns of ever-more-unlikely creatures allegedly encountered by them in the great forests—everything from furry trout and spiny-coated cactus cats to backward-flying birds. But the modern-day fascination with taxidermy jackalopes began much more recently—during the 1930s.

This was when Douglas Herrick, from Douglas, Wyoming (and later dubbed the "Father of the Jackalope"), had the bright idea of creating and then placing a stuffed jackalope in his shop's window. He and his brother Ralph, who helped to manufacture it, hoped that it would attract attention and customers, which it certainly did—so much so that jackalopes soon began appearing all across the American West, ultimately transforming this non-existent beast into a veritable icon. Today, marking the Herricks' integral contribution to the jackalope's success story, an 8-ft.-tall jackalope statue stands in the centre of Douglas, which is known (of course!) as Jackalope Square.

THE FACTS BEHIND THE FICTION?

Is it possible, however, that Eurasia's horned hares and North America's jackalopes are not entirely the product of folklore and fakery but actually have at least a little basis in fact? The first strong evidence that this might indeed be true came in 1937, with the publication of Volume 4 of Canadian writer-naturalist Ernest Thompson Seton's book *Lives of Game Animals*. In the section discussing cottontail rabbits, Seton included a page of self-drawn sketches of rabbits bearing a grotesque array of horn-like protrusions from their heads and faces. Some of these recalled the bestiary illustrations of horned hares and the folktale descriptions of jackalopes.

But what had caused these freakish horns to develop? Studying cottontails inflicted in this manner, medical biologist Richard E. Shope had discovered just a few years earlier that they had been infected by a specific virus (nowadays known as the Shope papilloma virus). This transforms facial follicle cells into hard tumors called papillomas, which in the most extreme cases give rise to these bizarre horns. One such specimen was found dead in a Minnesota woman's garden in September 2005 after she had called the police in alarm upon finding it there.

Shope also discovered a second virus that has similar effects, the Shope fibroma virus. Both are most common in North America. But, in Europe, hares are prone to a virus known as Leporipoxvirus, to which rabbits are also susceptible, and which again causes the production of horny facial nodules and growths.

It is easy to see how, centuries ago in pre-scientific times, if someone encountered a rabbit or hare in Eurasia or North America suffering from one of these viruses and bearing various horn-like growths on its head or face, belief in the existence of a rare, exotic species of horned hare or jackalope could swiftly develop. And as microbiological knowledge was at best inaccurate and at worst non-existent in those bygone ages, even naturalists might readily have been fooled into assuming that such horns were natural structures rather than the physical effects of a viral infection. Coupling this with the all-too-frequently exaggerated and distorted illustrations of animals present in bestiaries back then, and suddenly a hare with horns, or even a rabbit with antlers, is no longer so surprising and implausible a creature after all.

Chapter 4:
ALONG CAME A (GIANT) SPIDER...

Suddenly something was swinging between him and his terrible enemy—a thing like a large black crab, which was suspended at the end of a thin, glistening cord. It had appeared so swiftly, so silently, that for the moment Monty could not believe his eyes. He remained as still as a stone image, hardly breathing. The apparition was within a yard of his face, and exactly between him and the ape. As he stared at it, the thing spun half round and two glowing red specks caught his eyes.

In a flash he knew the truth.

This was a spider—a spider of a size that no naturalist had hitherto dreamt of. Yes, he saw it now, with its evil-looking head, and its long hairy legs curved in close to its body.

Christopher Beck, *The People of the Chasm*

Monstrous spiders of gargantuan size are perennially popular subjects in science fiction B movies as well as in classic fantasy novels such as J.R.R. Tolkien's *The Lord of the Rings* trilogy and *The Hobbit*, but could such beasts exist in reality? The answer routinely given by mainstream scientists to this question is a resounding "No!"—noting that the largest formally-recognized species of spider alive today is South America's goliath bird-eating spider *Theraphosa blondi*. (A related species, *T. apophysis*, the pinkfoot goliath bird-eater, sometimes has slightly longer legs but is less bulky.) The current record-holder for *T. blondi* is a 12-year-old captive female specimen called Rosi, which boasted a leg-span of 11.25 in (big enough to cover a dinner plate), a body weighing 6.17 oz—which is as heavy as six house sparrows—and as big as a tennis ball, plus a total body length of 4.75 in. Although these are impressive statistics, they are far from monstrous. In contrast, as revealed here in what is the most comprehensive coverage of giant cryptozoological spiders ever compiled, there are some remarkable yet currently-unresolved modern-day reports on file hinting that certain truly astonishing arachnids whose size very dramatically surpasses this

The author with a model of a giant spider (Karl Shuker)

latter species' stature lurk in shadowy zoological anonymity within various regions of our world.

A PUPPY-SIZED SPIDER IN PAPUA NEW GUINEA

The Kokoda Track (or Trail) is a predominantly straight, single-file foot thoroughfare running 60 miles through inhospitable terrain across the Owen Stanley mountain range of Papua New Guinea, and from July 1942 to January 1943 it was the site of a series of World War II battles between Australian and Japanese forces known as the Kokoda Track Campaign. In 2001, Australian cryptozoologists Peter and Debbie Hynes revealed that it was also here, while serving as a soldier in the Australian Army, that the father of one of Debbie's friends had a brief but unforgettable encounter with a mystery mega-spider:

> One day he had to take himself off into the scrub in answer to a call of nature. While thus engaged he noticed he was crouched down next to a very large cobweb—not the classic "fishing-net" sort but the fine, snow-white cottony stuff that spread all over the ground and tree trunk etc. His eye followed it one way and then the other—seems it was very extensive, like 10 to 15 ft either way. Then he noticed the spider itself, only a foot or so away from his face. It was a real horror—the body, i.e. thorax+abdomen, he described as the "size of a small dog or puppy", it was coloured jet black, the legs were thick and hairy but not as long as the classic "dinner plate tarantula" type spider that owes its size to the spread of its legs. This thing had more body bulk than spread. Needless to say he backed out of there *very* slowly and carefully.

In spiders, the "body" is actually just the abdomen (opisthosoma), not the thorax plus abdomen (although it can look like that to laymen unfamiliar with spider anatomy), because the thorax section is combined with the head, yielding the prosoma or cephalothorax. So, judging from the above description, the Papuan 'puppy spider' must have been at least the size of an adult chihuahua!

This is not the only report of a giant mystery spider encountered in New Guinea during World War II. During an interview with cryptozoologist Rob Morphy of AmericanMonsters.com on the U.S. radio show "Coast To Coast AM" a couple of years ago, a telephone caller named Craig recounted how his grandfather, while serving in New Guinea during WW2, encountered a monstrous spider in a web that scared him so much he hacked it to death with his machete.

According to Craig's grandfather, the spider measured an immense 3 ft. from tip to tip, and, unexpectedly, was not hairy like many big spiders are. Instead, it was shiny, and was emerald green in color. This nightmarish arachnid was encountered near Port Moresby, the capital of Papua New Guinea.

J'BA FOFI—"GREAT SPIDER" OF THE CONGO

Yet even this monster pales into insignificance alongside the horrifying j'ba fofi ("great spider") claimed by the Baka pygmy tribe and also the local Bantu hunters to exist amid the central African jungles of Cameroon and also the Democratic Republic of Congo (formerly the Belgian Congo). This eight-legged terror was first brought to attention in 2001, when cryptozoological explorer Bill Gibbons told me of a very frightening close encounter that had occurred one day back in 1938.

This was when explorers Reginald and Margurite Lloyd were driving along a jungle path in the Belgian Congo's interior. Suddenly, a figure stepped out onto the path just ahead of them, resembling a monkey or a small, stooped human. Reginald Lloyd stopped the car to let the figure pass and was astonished to see that it was a huge brown tarantula-like spider, with a leg-span of 3-4-ft! As he turned to grab his camera, however, the giant spider scuttled into the undergrowth and disappeared.

In November 2003, during an expedition to Cameroon seeking a mysterious long-necked water beast called the mokele-mbembe, Gibbons mentioned to the Baka pygmies the Lloyds' sighting (originally recounted to him by their daughter, Margaret). They were familiar with such creatures and provided him with additional information.

The Baka claimed that these colossal spiders were once quite common in this area but are rarer now (due to modern deforestation here?), although they had reputedly sighted one as recently as June 2003. These spiders used to construct hut-like lairs from leaves near to the pygmies' villages, and by spinning mighty webs between adjacent trees, with trip lines running across game trails, they ensnared and devoured victims as sizeable as duiker (small antelopes). Moreover, they were said by the Baka to be powerful and venomous enough to kill humans, too, but are themselves killed by the pygmies who encounter by them. The j'ba fofi supposedly lays white peanut-sized eggs, from which yellow spiderlings with purple opisthosomas emerge, turning brown as they mature.

SOUTH AMERICAN MEGA-SPIDERS

Reports of comparably massive spiders have also been recorded from the rainforests of Venezuela in South America. In 2008, the American television series "MonsterQuest" sent tarantula expert Rick C. West to investigate such stories in the field via a short, filmed expedition to some Venezuelan jungle villages near to the Orinoco River and the border with Colombia. During his three-day foray, he was accompanied by a team of local helpers and an experienced Amazon guide, Juan Carlos Ramirez, who has worked there for over 20 years.

West began his quest by visiting the village of San Rafael de Manuare. Here, one villager attested that as a child he had seen a giant tarantula-like spider capture a small dog from the village and drag it off into the jungle. Its opisthosoma was as big as a basketball, and when it reared up it was the size of a human. If such a gigantic spider existed, and its fangs (chelicerae) were in proportion to the rest of its body, they would probably measure 6-9 inches long. Although such claims would incite considerable scientific skepticism, Ramirez was convinced of the villagers' veracity, stating that they know the local fauna very well, and would not mistake something familiar, such as a monkey or a sloth on the ground, for a giant spider.

West and his team also visited Pandari, a village deeper in the mountains. Here, two inhabitants, Antonio and his son Simoni, spoke of a small child who had disappeared, never to return—which had been blamed upon giant spiders. In addition, so real is the Pandari villagers' fear of such creatures that they even engineer their huts specifically to keep them out, building thatched roofs that extend all the way down to the ground, thus yielding dense tightly interwoven barricades.

On the third day of West's expedition, they headed back into the jungle and found an extremely large spider lair in the ground, inside which they placed a videoscope. This revealed the presence of a very big tarantula, which they captured alive. Although nothing like as sizeable as the reputed chicken-killing, dog-devouring, child-abducting specimens feared by the villagers, it was roughly the same size as the biggest tarantulas on record and weighed 2 ounces inside a plastic specimen bag.

Sadly, West's expedition ended without finding conclusive evidence for Venezuela's fabled giant spiders. However, he was sufficiently impressed by the size of their captured spider to consider it possible that bigger ones did exist in the jungle, and stated that he planned on returning to continue the search for one.

In 2011, British cinematographer Richard Terry sought horse-killing, child-abducting giant spiders in Colombia's rainforest, for the television series "Man v Monster." He didn't find any either, but villagers claimed that these dreaded beasts inhabited subterranean lairs opening onto the forest floor via huge holes.

MONSTER SPIDERS IN VIETNAM

On April 8, 2013, American cryptozoologist Craig Woolheater posted on the Cryptomundo website a fascinating communication that he'd received from an American correspondent publicly identified only by his Cryptomundo user name mrmaxima. This person stated that their father-in-law claims that while serving in the jungles of Vietnam during the Vietnam War as part of a five-man unit conducting scout work, he encountered spiders with bodies the size of dinner plates, and, with their legs, yielding a total span of 20-30 inches. These terrifying arachnids were always spied near creeks or other water sources, and were so tough that even after being shot with M16s and unloaded full magazines, they were still moving around.

THE HOBBIT ACTOR AND A REAL-LIFE SHELOB?

British actor Dominic Monaghan has starred in many films and TV series, but is probably best known for his role as the hobbit Merry in Peter Jackson's spectacular *The Lord of the Rings* (*LOTR*) movie trilogy, based upon J.R.R. Tolkien's monumental fantasy epic. In 2011, however, I came upon a very intriguing online report previously unknown to me that suggested Dominic might have been taking his *LOTR* role even more seriously than expected.

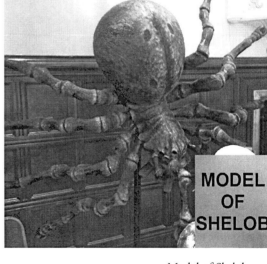

Model of Shelob from *The Lord of the Rings* (Karl Shuker)

In *LOTR*, two of Merry's fellow hobbits, Frodo and Sam, experience a life-threatening encounter with Shelob—a colossal spider. Yet whereas Shelob was fictional, the media report, which was reproduced on several websites, claimed that Dominic was planning to launch a 12-man expedition in search of a real-life mega-spider that, if proven to exist, would be the world's largest.

According to this report (formerly at http://www.inoutstar.com/news/In-Search-For-The-8-Legged-Freak-5443.html), which was written by Alexandru Stan and published on InOut Star's website on March 12, 2008, the spider in question was called the hercules baboon spider, measured 14-15 inches across, and, of particular interest, was known only from a single specimen obtained in Nigeria during the early 1900s, which is now preserved in alcohol at London's Natural History Museum.

As the Life Sciences Consultant for *Guinness World Records* (previously known as *The Guinness Book of Records*), I am naturally well aware that the current record-holder as the world's largest species of spider by mass is, as noted earlier, the goliath bird-eating spider *Theraphosa blondi*. This nocturnal burrowing species, native to wet swamps and marshy areas in the mountain rainforests of northeastern South America, boasts a leg span of up to 12 in (the diameter of a dinner plate!), and can weigh over 6 oz. Consequently, any spider that allegedly exceeds these dramatic dimensions, and, as a bonus, is also virtually unknown to science, is definitely going to attract my undivided attention!

Eager to learn more, in case Guinness's existing record for the largest spider species by mass needed to be amended, I lost no time in researching this subject personally and also contacting a number of likely sources of further information, including the Natural History Museum itself. Over the next few weeks, a welter of data came my way, the most informative of which can be summarized as follows.

An adult female *Theraphosa blondi* (public domain)

A BEWILDERMENT OF BABOON SPIDERS

Firstly, I discovered to my surprise that in spite of the report's claim that only a single specimen of hercules baboon spider existed, there seemed to be a veritable embarrassment of specimens in the pet trade. There was also a plentiful supply of videos posted on *YouTube* of what were claimed to be hercules baboon spiders. Something, clearly, was amiss here—but what?

After learning of my desire to discover as much as I could regarding this enigmatic species, palaeontologist Darren Naish, who shares with me an interest in cryptozoological matters, sent me the following illuminating email on March 16, 2011:

> It is implicated by spider people that the things now being called "hercules baboon spiders" are actually nothing of the sort (they're actually king baboon spiders), and that the animal that really should go with this name is indeed only known from the type specimen.

Pursuing this promising lead, I discovered that the "real" hercules baboon spider is *Hysterocrates hercules*, which is genuinely known only from its type specimen—a female collected by a Lieutenant Abadie at Jebba in what was then Upper Niger, now Nigeria. Characterized by a black cuticle covered with a thick coating of dark olive-brown hairs, and shining with a grayish silky sheen under reflected light, as well as by having the fourth leg unthickened, this unique specimen is held at the Natural History Museum, and its species was formally described in the November 14, 1899, issue of the *Proceedings of the Zoological Society of London* by none other than renowned British zoologist Reginald I. Pocock.

As for the so-called hercules baboon spider specimens in the pet trade and on *YouTube*, these pretenders to the throne of *H. hercules* are indeed specimens of the related but smaller king baboon spider *Pelinobius muticus* (aka *Citharischius crawshayi*, a junior synonym), which is an East African species with a leg span of up to 8 in. So the first riddle concerning the hercules baboon spider was duly solved.

On April 21, 2011, I received another very insightful email, this time from Richard Gallon, the administrator of the British Tarantula Society Study Group:

> I can confidently state that Pocock's holotype specimen of *Hysterocrates hercules* (which I have measured and examined for forthcoming taxonomic papers) does not even come close to members of the genus *Theraphosa* in terms of leg-span or body mass.
>
> *Pelinobius muticus* females are large and bulky, but not as large as *H. hercules* or any *Theraphosa*. Indeed there are several South American genera (e.g. *Pamphobeteus, Sericopelma, Lasiodora* etc.) which are bigger than these African taxa.

Judging from this, Dominic's notion that *H. hercules* may well be the world's biggest spider appeared to be in error—an assumption comprehensively confirmed when I received on May 21, 2011 an email from George Beccaloni of the Natural History Museum containing the following crucial details:

> With regard to the West African baboon spider *Hysterocrates hercules,* it is certainly very large, but the heaviest spider known is undoubtedly the 175 gram [6.17 oz] female *Theraphosa blondi* listed by Guinness, and the largest in terms of legspan is the giant huntsman *Heteropoda maxima,* from caves in Laos, which has an accurately measured maximum legspan of 300 mm [12 in]...Any claims of larger spider specimens remain to be proven—and I don't think that any baboon spider is likely to displace the current champions!

On June 3rd, Jan Beccaloni, the museum's arachnid curator, also emailed me, kindly enclosing some information received from fellow arachnologist and theraphosid specialist Ray Gabriel from the British Tarantula Society, who stated that *H. hercules* is only about two-thirds the size of *T. blondi.*

So there it was—everyone was agreed that *H. hercules* was neither the heaviest species of spider nor even the species with the greatest leg span. And that seemed to be the end of the matter, until...

CLASH OF THE TITANS!

In a subsequent email, Beccaloni suggested a highly entertaining means of publicizing this former controversy—by staging at the Natural History Museum a filmed comparison by volume of the type specimen of *H. hercules* with a suitably sizeable specimen of *T. blondi*, supervised by a member of the *Guinness World Records* (*GWR*) team. Needless to say, I considered this to be an excellent idea, and passed it on at once to the editor of *GWR*, Craig Glenday, who thought so too, and was happy to act personally in the capacity of official adjudicator.

And so it was that my original investigation of *H. hercules* ultimately led to a filmed "Battle of the Spiders" weigh-in at the Natural History Museum in June 2011, conducted by George Beccaloni and witnessed by Craig Glenday for *GWR*. Utilising Archimedes' Principle of liquid displacement as a means of accurately determining the volume of the challenger (the *H. hercules* holotype) and the defender (a hefty adult female *T. blondi* called Tracy, a long-deceased pet of Jan Beccaloni) as both were preserved in alcohol, the two species' rival claims to the title of the world's heaviest spider were finally put to the test.

The Natural History Museum's online press release contained a video of this historic arachnological bout (which can be viewed at http://www.nhm.ac.uk/about-us/news/2011/july/worlds-heaviest-spider-title-challenged-at-museum99065.html).

And the result? Tracy's volume was more than double the *H. hercules* type specimen's. In short, a straight knockout, with *T. blondi* the undisputed heavyweight spider champion, retaining its title with ease.

Even so, as a nonetheless respectably (albeit not superlatively) large species yet still known only from a single specimen, *H. hercules* retains an air of mystique. Moreover, so far Dominic Monaghan has not launched an expedition to look for it. Consequently, an excellent way to bring this investigation to a satisfactory close would be for an intrepid spider seeker to pursue the hercules baboon spider in the field, as the rediscovery of this long-lost semi-Shelob is certainly long overdue!

GIANT SPIDERS IN SUBURBIA

One of the most startling giant spider reports comes from Leesville in Louisiana. According to William Slaydon, it was here, while walking northwards along Highway 171 to church one cool night in 1948, that he, his wife, and their three young grandsons had spied a gigantic spider—hairy, black, and memorably described as "the size of a washtub." It emerged from a ditch just ahead of them and crossed the road before disappearing into some brush on the other side. Not surprisingly, the family never again walked along that particular route to church at night!

Nor is that the only report of a giant spider in suburbia. On February 11, 2013, Adam Bird from Nottingham, England, shared the following remarkable, never-before-publicized account on Facebook. It was related to him by a local librarian, Sheila, who had encountered the spider in question about 12 years earlier. One evening, Sheila was driving along Nottingham's Stone Bridge Road, on one side of which was a farm (still there today) and on the other side a disused factory (now demolished). As she approached the factory, her car's headlights lit up what she thought at first was a hedgehog, crawling towards the factory site. As she drove nearer, however, she realized to her horror that it was a huge, hairy, tarantula-like spider. Sheila estimated that its body alone was the size of a large dinner plate, and when adding in the length of its legs, she deemed its total width to be about 2 ft. She continued to watch as it scuttled across the road and through the fence into the factory, then she quickly drove away, but, not surprisingly, the memory of this spine-chilling encounter has remained with her ever since.

SUSPECT SOLIFUGIDS AND WHIMSICAL WITCH SPIDERS

In the spring of 2004, during the military conflict in Iraq, a photo of a supposedly monstrous spider captured in the desert there by an American soldier was widely circulated on the internet. When I examined it, however, I recognized straight away that the "creature" it portrayed was in fact *two* solifugids or camel spiders, one clamped via its huge jaws to the abdomen of the other one, and photographed at such an angle as to appear far bigger than they truly were, i.e. a classic example of forced perspective. Although related to true spiders, solifugids constitute a separate taxononic order of arachnids.

A photograph that has been doing the rounds of websites online for quite a while depicts what looks like an absolutely colossal spider upon the side of a house. According to the blurb that often accompanies this image, it is an Angolan witch spider, an immense species that supposedly migrated into Texas from South America (despite being called Angolan??), preys upon cats and dogs, and takes several bullets to kill it. In reality, however, as exposed by Snopes.com, a famous rumor-debunking website, the photo is merely a digitally modified version of one depicting a normal-sized wolf spider, snapped by artist Paul Santa Maria.

AN EIGHT-LEGGED SIREN FROM FRANCE

Undoubtedly the least likely of all giant spider reports, however, featured a veritable siren, albeit one with eight legs instead of a fishtail, yet equally as gifted musically (not to mention homicidally) as any of Greek mythology's lethal mermaids.

The earliest documentation of it that I have so far uncovered is a detailed account in the *Ann Arbor Argus* newspaper for September 14, 1894, which was subsequently reprinted in various other American newspapers. According to this extraordinary journalistic concoction of Grand Guignol and cryptozoology, during late March (and invariably at night) each year for many years, men and women had inexplicably been disappearing in a region of Paris known as the Tomb of Issoire, without any trace of them ever being discovered. One night, however, a policeman in this vicinity happened to hear a strange musical song issuing forth from a hole at the base of a huge rock, dubbed the Giant's Cave due to the legend that a giant had been buried there long ago.

As the policeman stood listening, he saw a young man approach the hole, seemingly hypnotized by the unearthly strains issuing forth from their hidden subterranean source, and then suddenly the man raced into the hole at full speed. The policeman chased after him, firing his revolver to alert some of his colleagues for backup as he entered the hole. They soon arrived, by which time the strange music had ceased, its mysterious melody having been replaced by the sounds of a violent struggle. Arming themselves with ropes, ladders, and lamps, the policeman's colleagues swiftly penetrated the chasm into which the hole led, and beheld a terrifying sight.

The young man, dead, was in the grip of a monstrous spider, which according to the newspaper report was:

> ...as large as a full grown terrier, covered with wartlike protuberances and bristling with coarse brownish hair. Eight jointed legs, terminated by formidable claws, were buried in the body of the unfortunate victim. The face had already disappeared. Nothing could be seen but the top of the head, and the monster was now engaged in tearing and sucking the blood from his throat.

Several blasts from the policemen's guns soon dispatched this horrific creature, after which they found their colleague lying in a corner, unconscious but unharmed. They then carried the two men and the carcass of the great spider back out through the hole into the Parisian street where this surreal incident had begun.

In an even more bizarre twist to this already exceedingly weird tale, however, the newspaper report ended by claiming that the spider belonged to a species hitherto deemed extinct for centuries, named as *Arachne gigans*, which supposedly possessed the ability to entice its victims by giving voice to a mesmerizing song, and which may be represented by additional live specimens still existing in the deepest galleries of this city's catacombs. But what happened to the slain Issoire mega-spider? According to the report:

> The dead body of the spider was conveyed to the Museum of Natural History, where it was carefully prepared and stuffed and is now on exhibition.—Once a Week.

The museum was presumably France's National Museum of Natural History, in Paris, but it will come as no surprise to discover that the museum has no record of ever even receiving, let alone exhibiting, such a singular specimen. Nor is there any such species as *Arachne gigans*, living or extinct.

During the late 19th century and early 20th century, it was commonplace for newspaper editors to spice up their publications with lurid tales of extraordinary discoveries and events that had no basis in reality. And certainly, there seems very little doubt that the giant singing spider of Issoire only ever existed within the fevered imagination of an editor anxious to fill up some spare column inches inside his newspaper.

PHYSIOLOGICAL SIZE LIMITATIONS

But what about the other monsters reported here—could immense spiders truly exist? Other than Leesville and Nottingham, the areas where they have been reported are all sufficiently impenetrable, inhospitable, and little-explored to be potentially capable of hiding some notable zoological surprises. However, the fundamental problem when considering giant spiders is not one of zoogeography but rather one of physiology. Their tracheal respiratory system (consisting of a network of minute tubes carrying oxygen to every cell in the body) prevents insects from attaining huge sizes in the modern world, because the tracheae could not transport oxygen efficiently enough inside insects of giant stature. During the late Carboniferous and early Permian Periods, 300 million years ago, huge dragonflies existed, but back in those primeval ages the atmosphere's oxygen level was far greater than it is today, thereby compensating for the tracheal system's inefficiency.

Some of the largest known spiders also utilize a tracheal respiratory system, whereas smaller spiders employ flattened organs of passive respiration called book lungs. Yet neither system is sufficiently competent to enable spiders to attain enormous sizes, based upon current knowledge at least. So if a giant spider does thrive in some secluded, far-off realm, it must have evolved a radically different, much more advanced respiratory system, not just a greatly enlarged body.

All of this, however, is sheer speculation, and is likely to remain so—unless, for instance, in the not-too-distant future a Baka pygmy should happen not only to kill a j'ba fofi but also to preserve its body afterwards, and duly alert scientific attention to it. Then at last we might have the long-awaited solution to this fascinating mystery—although arachnophobes might be more than happy for it to remain unsolved indefinitely!

Chapter 5:
NEVER TANGLE WITH A TYGOMELIA— SEEKING FORGOTTEN MYSTERY BEASTS FROM BYGONE TIMES

> *...the Indians may not always be wrong in their beliefs, or in the tales they tell of strange and wonderful animals or reptiles that still lurk in the forest depths. I am not, of course, maintaining that all their wild tales and fantastic beliefs are true; I only suggest that here and there, there may be a better foundation for them than many, perhaps, would think. I have myself seen strange creatures that are entirely unknown to zoologists, but they got away before I could shoot or catch them.*
>
> — Frank Aubrey, "The Spell of the Bird"

For every Nessie, bigfoot, Beast of Exmoor, yeti, or Mongolian death worm, there is a veritable host of other, far less familiar mystery beasts on record—elusive creatures that have been all but forgotten even by the cryptozoological cognoscenti, let alone by mainstream zoologists. In this chapter, however, I aim to redress the balance somewhat by introducing, or reintroducing, a diverse selection of remarkable but scarcely-publicized, scientifically-unrecognized animals from bygone times that hitherto have slipped through the cracks between the paving stones of cryptozoological prominence and descended ignominiously into the dank, unlit realms of public obscurity.

TYGOMELIA—A GIRAFFE IN MOOSE'S ANTLERS?

One of the most incredible and perplexing cryptids on file made its media debut in an *Ottawa Times* newspaper article of November 22, 1870, after which it vanished from the headlines as swiftly as it had entered them:

> Sir John E. Packenham, an officer in the English army, who has been spending the last year in her Majesty's northern provinces, arrived at Fort Buford [in North Dakota] with an animal of rare beauty, and never before caught on this continent, nor has it been known till late years that the species existed in this country. It is of the same family as the giraffe, or camelopard, of Africa, and is known to naturalists as the tygomelia. They are known to inhabit the high table lands of Cashmere and Hindoo Kush, but are more frequently seen on the high peaks of the Himalaya Mountains. The animal was taken when quite young, and is thoroughly domesticated, and follows its keeper like a dog. It is only four months old, and ordinarily stands about five feet high, but is capable of raising its head two feet, which makes the

animal seven feet when standing erect. It is of a dark brown mouse color, large projecting eyes, with slight indications of horns growing out. The wonderful animal was caught north of Lake Athabasca, on the water of the McKenzie's River. It has a craw similar to the pelican, by which means it can carry subsistence for several days. It was very fleet, being able to outfoot the fastest horse in the country. The black dapper spots on the rich brown color make it one of the most beautiful animals in existence, more beautiful than the leopard of the Chinese jungle. Sir John did not consider it safe to transport this pet by water down the Mississippi River, fearing the uncertain navigation and the great change of climate from the Manitoba to the sunny south. He has, therefore, wisely concluded to go by way of St. Paul, Minnesota. The commander of Fort Buford furnishes him with an escort for the trip. He will then proceed through Canada to Montreal, where he will ship his cargo to England.

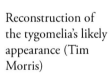

In reality, however, no such beast is known from India or North America—or, indeed, from anywhere else!

Moreover, the only plausible suggestion regarding its identity that has been offered to date (always assuming, of course, that the original report was not a journalistic spoof) is one suggested by Edmonton-based cryptid investigator Kevin Stewart. Namely, that this "tygomelia" was a young, freakishly mottled moose *Alces alces* (such specimens have occasionally been documented in the wild). But with no images or further accounts of it known to exist, there is little likelihood that we shall ever know for sure.

Reconstruction of the tygomelia's likely appearance (Tim Morris)

STELLER'S SEA-BEAR—A POLAR BEAR IN JAPAN?

Georg Wilhelm Steller (1709-1746) was a German naturalist and explorer who took part in Vitus Bering's Second Kamchatka Expedition to Russia's Far East during the early 1740s, and whose greatest contribution to science was the discovery and description of several major new species of animal while the expedition was shipwrecked on Bering Island. These included a new jay, eider duck, sea eagle, sea lion, and, most famously, a gigantic sirenian, Steller's sea cow *Hydrodamalis gigas*, which, tragically, was hunted into extinction within the next three decades.

Steller also documented three very puzzling creatures whose identities remain unknown to this day. One was an extraordinary sea mammal variously likened to a seal or even a merman, which has been dubbed Steller's sea ape and which I documented in my book *Mysteries of Planet Earth* (1999). The second was a bird, Steller's sea raven, which I have chronicled in various articles. As far as I am aware, however, the third member of this mystifying trio is scarcely even remembered nowadays, has never previously been documented in any cryptozoology book, and is known as Steller's sea bear.

Steller recorded it as follows in his book *De Bestiis Marinis* [*Beasts of the Sea*] (1751):

> Report, as I gather from the account of the people, has declared that the sea bear, as it is called by the Rutheni and other people, is different. They say it is an amphibious sea beast very like a bear, but very fierce, both on land and in the water. They told likewise, that in the year 1736 it had overturned a boat and torn two men to pieces; that they were very much alarmed when they heard the sound of its voice, which was like the growl of a bear, and that they fled from their chase of the otter and seals on the sea and hastened back to land. They say that it is covered with white fur; that it lives near the Kuril Islands, and is more numerous toward Japan; that here it is seldom seen. I myself do not know how far to believe this report, for no one has ever seen one, either slain or cast up dead upon the shore.

Needless to say, this description readily calls to mind the polar bear *Ursus maritimus*—but there is one major problem when attempting to reconcile Steller's sea bear with the latter species. There is no confirmed record of the polar bear ever having been found anywhere near as far south as the Kuril Islands, let alone towards those of Japan.

Could it be, as suggested back in 2002 by Steller scholar Chris Orrick, that reports of polar bears seen further north (this species was already well known in Steller's time) had somehow become confused with those of fur seals, which are not white but are often loosely if inaccurately dubbed "sea bears" and are indeed found around the Kurils and Japan? As for his treating the very existence of the sea bear with suspicion, perhaps Steller was not aware of the polar bear's marine behavior, mistakenly assuming that it was exclusively terrestrial.

Alternatively, as opined by American cryptid chronicler Dale

Drinnon, perhaps reports of Steller's sea bear were based upon a relict southern population of polar bears left over in the Kuril-Japan area from a colder, earlier time when this species was more widespread but which was already dying out by the time that Steller had heard stories about such creatures here. This possibility has a notable precedent with Steller's sea cow, known from fossil evidence to have existed formerly over a very considerable area stretching from Japan to California but which had become confined to the Bering Strait by the time of Steller's arrival.

A SEA SNAIL WITH ANTLERS...AND PAWS!

It's amazing what you find when looking for something completely different. Browsing through some early books recently in search of some relevant illustrations for another writing project, I chanced upon the following astonishing beast from a medieval bestiary cum medical encyclopaedia entitled *On Monsters and Marvels*, written by Ambroise Paré, a 16th-century French surgeon.

Citing as his source of information the twentieth volume of *Cosmographie Universelle*—authored by one of his contemporaries, André de Thévet (1516-1590), a French Franciscan priest who was also a writer, explorer, and cosmographer—Paré reported that the Sarmatian or Eastern Germanic Sea (early names for the Baltic Sea) is home to a truly monstrous species of sea snail.

As thick as a wine cask, this singular and very sizeable creature is instantly distinguished from all other such mollusks by virtue of its pair of deer-like antlers. These decidedly un-snail-like accoutrements are borne upon the upper region of its head, and at the tip of each branch on each antler is a small, round, lustrous bulb resembling a fine pearl. By contrast, this snail's eyes, which in less exotic gastropods can be found at the tips of a pair of optic tentacles, are laterally sited on its head, just like those of many vertebrates, and glow brightly like candles.

Equally unexpected was Paré's claim that this very odd creature sports a roundish nose reminiscent of a cat's, complete with whisker-like hair all around it, plus a wide slit-like mouth beneath which hangs a fleshy projection of hideous appearance.

And as if antlers and a feline nose are not sufficiently bizarre characteristics for a sea snail (or, indeed, any other type of snail) to boast, eschewing the usual monopodial mode of locomotion common to normal gastropods this extraordinary marine mollusk possesses four fully-differentiated limbs, each with its own wide, hooked paw. It also sports a fairly lengthy tail, bearing a varicolored tigerine pattern. Moreover, the image accompanying this morphological description in Paré's book (and now reproduced here) shows the Sarmatian Sea's antlered mega-snail to bear a very large and sturdy, heavily annulated, whorled shell upon its back.

This remarkable creature apparently spends much of its time out at high sea on account of its timid nature but is sufficiently amphibious to be able to venture forth onto the seashore during fine weather in order to graze upon any marine plant life present there. An edible species itself, its flesh is said to be very delicate and tasty to eat, and its blood reputedly has medical properties, ameliorating sufferers afflicted with leprosy.

No snail matching the description communicated by Paré from Thévet's work is known to modern science. So could the giant antlered snail of the Sarmatian Sea be as mythical as the camphurch (a web-footed unicorn), the mer-folk, the winged unipodal humanoids, and certain other unquestionably fabulous entities also documented by these authors, or is it merely a somewhat distorted description of some bona fide animal?

Illustration of the Sarmatian Sea's antlered sea snail from Paré's 16th-century book (public domain)

Reading it through, I was reminded somewhat of the nudibranchs or sea slugs, many of which are extremely flamboyant in appearance, with all manner of ornate, plume-like embellishments known as cerata arising dorsally and laterally from their body, which may conceivably explain the "limbs" and "paws" reported for this creature. Nudibranchs also have a pair of long cephalic (head-borne) tentacles, and if a species existed whose tentacles bore projections they may resemble antlers. Moreover, the two eyes of nudibranchs are sited directly on their body, just behind the head, not on optic tentacles like those of snails, so a pair of laterally sited eyes would not be impossible. And nudibranchs are well documented from the Baltic Sea.

However, such an identity is instantly wrecked by the Sarmatian antlered snail's hefty shell, because nudibranchs are shell-less. In addition, all known nudibranchs are carnivorous, not herbivorous. And even the largest known nudibranchs do not exceed 2 ft. long—a far cry from any gastropod as thick as a wine cask.

So what is the likeliest identity of this mystery mollusk? Might it possibly have been an unknown species that became extinct before modern science was ever able to confirm its reality and add it to the zoological catalogue of formally-recognized lifeforms? Or was such an incredible creature as this never anything more than a wholly imaginary beast confined to the pages of early travelogues and compendia of monsters? To be, or not to be? That, indeed, is the question with this cryptid.

ZEBRO—AN EQUINE ENIGMA FROM IBERIA

During the Middle Ages and the Renaissance, several Spanish hunting treatises alluded to a mysterious, now-vanished equine creature known as the zebro (or encebro, in Aragon), living wild in the Iberian Peninsula. In one of these works, it was described as "an animal resembling a mare, of grey colour with a black band running along the spine and a dark muzzle." Others likened it to a donkey but louder, stronger, and much faster, with a notable temper, and whose hair was streaked with grey and white on its back and legs. What could it have been?

Although largely forgotten nowadays, the zebro experienced a brief revival of interest from science in 1992. That was when archaeologists Carlos Nores Quesada and Corina Liesau Vonlettow-Vorbeck published a very thought-provoking article in the Spanish periodical *Archaeofauna*, in which they boldly proposed that the zebro may have been one and the same as an equally enigmatic fossil species—*Equus hydruntinus*, the European wild ass.

The precise taxonomic affinities of this latter equid have yet to be satisfactorily resolved. Although genetic and morphological analyses suggest that it was very closely related to the onager *E. hemionus* (one of several species of Asiatic wild ass), it can apparently be differentiated from these and also from African wild asses by way of its distinctive molars and its relatively short nares (nasal passages). Arising during the mid-Pleistocene epoch, approximately 300,000 years Before Present, the European wild ass persisted into the early Holocene before finally becoming extinct. During the late Pleistocene, its zoogeographical distribution in western Eurasia stretched from Iran in the Middle East into much of Europe, reaching as far north as Germany, and it was particularly abundant along the Mediterranean, with fossil remains having been recovered from Turkey, Sicily, Spain, Portugal, and France.

According to Quesada and Vonlettow-Vorbeck, moreover, this species may have survived in southernmost Spain and certain remote parts of Portugal until as late as the 16th century (they consider its disappearance to represent the Iberian Peninsula's last megafaunal extinction), where, they suggest, it became known locally as the zebro. More recently, their theory gained support from the discovery of *E. hydruntinus* remains at Cerro de la Virgen, Granada, dating from as late as the 9th century.

Some researchers have also suggested that before dying out, the zebro gave rise at least in part to a primitive, nowadays-endangered Iberian breed of donkey-like domestic horse called the sorraia (which was once itself referred to as the zebro). Furthermore, many believe that it was from the term "zebro" that "zebra" originated as the almost universally used common name for Africa's familiar striped equids.

Even today, many Iberian place-names still exist in which the mysterious but now-obscure zebro's name is preserved. These include Ribeira do Zebro in Portugal; and Valdencebro (in Teruel), Cebreros (in Ávila), Encebras (Alicante), and Las Encebras (Murcia) in Spain.

THE WHALE-FISH OF LAKE MYLLESJÖN

Since at least the mid-1800s, rumors have been circulating of an enormous water beast dubbed the whale-fish that is said to inhabit the cool, deep waters of Lake Myllesjön in southernmost Sweden. Yet were it not for Swedish cryptozoological artist Richard Svensson, it is likely that this freshwater cryptid would have remained wholly unknown beyond his country's perimeters. Happily, however, Richard permitted me to include information he had gathered as well as a detailed line drawing he had prepared for my book *Mysteries of Planet Earth* (1999), which brought it at last to international attention. More recently, Richard has also very kindly sent me a whale-fish model he had designed and produced, which is depicted here for the very first time in any English-language book.

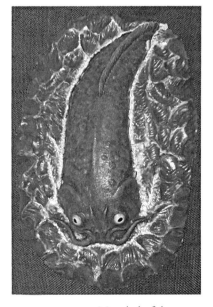

My whale-fish model, produced by Richard Svensson (Karl Shuker)

A flurry of sightings occurred during the 1920s and 1930s, including one featuring a huge fin-crested back rising above the water surface. Another eyewitness claimed to have seen a whale-like creature frolicking at the water's edge, and local fisherman Sven Johan discovered two of his fishing nets torn apart after hauling them out of the lake. Moreover, three girls fled screaming in terror after seeing what they likened to a whale basking in shallow water near the lake's shore.

Attempts to capture the monster failed, however, as they also did in the 1960s following several stories hitting the Swedish headlines of huge "logs" splashing in the lake. Richard has learned of reports as late as the 1970s, but when he visited the lake in 1996, he was horrified to discover that a sizeable motorway had been constructed right next to it, and that people living nearby were no longer familiar with reports of the whale-fish. This suggests that even if such a creature had indeed existed here in the past, it no longer does so, possibly extirpated during the motorway's construction.

Richard and I agree that the apparently erstwhile whale-fish of Lake Myllesjön (and possibly also of another Swedish lake once said to harbor such a creature) was probably a giant European catfish (wels) *Silurus glanis*. Richard states that the largest confirmed wels specimen ever caught in Sweden, way back in 1871, measured a mighty 11.75 ft. long.

WHITHER THE WATER ELEPHANT?

During the first decade of the 20th century, French traveler-explorer M. Le Petit twice encountered a truly bizarre form of elephant in the Congo region of central Africa. The first was merely a brief sighting, at a considerable distance, of a creature said by locals to be a water elephant, seem swimming in the River Congo with its head and neck above the water surface. The second encounter, however, was much more memorable, as Le Petit allegedly witnessed no less than five such creatures, this time on land and at close range, amid tall grass in a swampy area sandwiched between Lakes Mai-Ndombe and Tumba, within the borders of what is now known as the Democratic Republic of Congo.

Le Petit claimed that they each stood 6.5-8.25 ft. at the shoulder, with an elongate neck, curved back, smooth and shiny hairless hippopotamus-like skin (but darker in color), relatively small elephant-like ears, and fairly short legs whose feet each sported four toes. Most

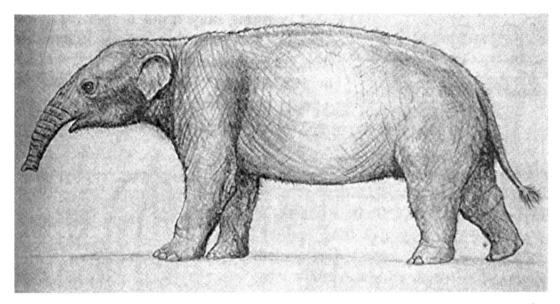

Reconstruction of a water elephant (Markus Bühler)

distinctive of all, however, was the noticeably elongate head, ovoid in shape, with only a very short trunk and no tusks (thus resembling the head of a giant tapir).

Le Petit shot one of them in the shoulder, but was unable to procure the wounded creature, after which all five retreated into deeper water and vanished from sight. According to native Babuma hunters, who refer to this unclassified species as the ndgoko na maiji, it spends the entire day in deep water, only emerging onto land at night, where it grazes on rank grass, and is not common.

Nothing more was heard of these peculiar aquatic pachyderms for many years—not until July 2002, in fact. That was when veteran field cryptozoologist Bill Gibbons informed me that the aviators inside a military helicopter that had recently flown over Lake Tumba claimed to have seen a herd of very odd-looking elephants fitting Le Petit's description of the water elephants seen all those decades earlier. Sadly, however, the aviators' sighting was not pursued, so the identity of their mystery beasts remains undiscovered—just like the water elephant itself.

That is a great tragedy because what makes this cryptid so intriguing is that its description is remarkably reminiscent of certain primeval pachyderms dating back many millions of years, to the very base of the proboscidean family tree. Indeed, the "dawn elephant" itself, *Moeritherium*, though smaller in size, not only resembled the water elephant in general form but also is believed to have been at least partially aquatic. Moreover, if this archaic elephant had given rise to a lineage of larger but otherwise similar beasts (in both morphology and lifestyle), surviving unchanged to the present day within remote, largely inaccessible swamplands, the result would have been a modern-day species readily recalling the Congolese water elephant. Just a coincidence?

THE TAILED SLOW LORISES OF ASSAM'S LUSHAI HILLS

Over a century has passed since they were reported, yet Assam's tailed slow lorises remain an enduring primatological mystery.

The only known photograph of tailed slow lorises from Assam's Lushai Hills (public domain)

Two of these white-coated, woolly-furred prosimians were captured and photographed by a T.D. La Touche of the Geological Survey of India during December 1889 (though not documented until 1908), in the jungle near Fort Lungleh in Assam's Lushai Hills, during the Lushai Expedition of 1889-90. However, they escaped shortly afterwards and were not recaptured. Overall, they seemed akin to the familiar slow lorises of the genus *Nycticebus*, as they possessed short but stout limbs, a large rounded head, flat face and small muzzle, short roundish ears, large eyes each encircled by a dark triangular patch, and a narrow black stripe running from the skull's occipital region along the entire length of the back.

However, as clearly revealed in the unique photograph reproduced here, which originally appeared in a short article by Nelson Annandale, Superintendent of the Indian Museum, who documented them in the *Proceedings of the Zoological Society of London* (*PZSL*) for November 17, 1908, the remarkable loris form represented by these two briefly-captive specimens differed dramatically from all other lorises in one very conspicuous way—it had a thick bushy tail!

In stark contrast, the tail of all *known* species of slow loris is vestigial.

Moreover, confirming that the tail was not simply an optical illusion within the photograph, La Touche assured Annandale that it was indeed present in both of the captive animals. And in a postscript of January 5, 1909 to his *PZSL* article, Annandale announced that he had learned from a Colonel E.W. Loch that the tailed slow loris of the Lushai Hills was well known to him, too.

Unless it is a teratological, freak variety of the Indian slow loris *N. bengalensis*, the tailed slow loris of Lushai Hills must surely constitute a radically new species still awaiting official recognition. Consequently, in my book *The Lost Ark: New and Rediscovered Animals of the 20th Century* (1993), I proposed *Nycticebus caudatus* ("tailed slow loris") as a formal scientific name for it.

Perhaps the greatest riddle of all, however, is why such a visually distinct form of loris has apparently never again been reported since the 1909 postscript to Annandale's *PZSL* account, not even by local Indian naturalists who would surely have been fascinated by such a creature. Judging from Col. Loch's statement that it was well known to him, the two captured specimens were not the only ones that existed back in the early 20th century, so what has happened since then—has this unique form died out? Clearly, the mystery of the tailed slow lorises of the Lushai Hills endures, even if, tragically, the lorises themselves no longer do.

FORGET JAWS, HERE'S GUMS!

One of the world's most famous explorers is, or was, Lieutenant-Colonel Percy Fawcett, who mysteriously vanished in 1925 while exploring the vast uncharted jungles of Brazil. Before his tragic disappearance, however, he had penned a fascinating account of his explorations in these lands, which was published in 1953 as *Exploration Fawcett*. It contains brief accounts of several mystifying creatures, but none more so than a supposed giant toothless shark.

According to Fawcett, the Paraguay River contains "a freshwater shark, huge but toothless, said to attack men and swallow them if it gets a chance." In reality, however, very few sharks inhabit freshwater, and those that do are far from toothless. So if such a shark really does exist, it is dramatically different from anything currently known to science. However, it may not be a shark at all. British cryptozoological researcher Mike Grayson has opined that it could be a very large sturgeon, some species of which are vaguely shark-like and can attain great sizes. However, he concedes that this is a very tentative identification, acknowledging that sturgeons do not attack and swallow people, and that there is no known species of South American sturgeon to act as a zoogeographical precedent anyway.

My own feeling is that a catfish identity may be more plausible. There are more species of catfish in South America than anywhere else, and some of these latter are among the world's largest, too. In Europe, the giant, earlier-noted wels catfish has often been accused (unjustly or otherwise) of swallowing people, but even if this aspect is folkloric, the concept of a giant catfish existing in South America is by no means impossible.

True, catfish usually possess teeth at least on their vomer bone. However, old specimens of some species are entirely toothless, as in, significantly, the giant pa beuk *Pangasianodon gigas* of southern Asia's Mekong River. Up to 10.5 ft. long, it is the world's largest species of fish confined entirely to freshwater, yet amazingly it remained undiscovered by science until as late as 1930. Perhaps Fawcett's giant toothless so-called shark— aptly dubbed "Gums" by Grayson—is in reality an elderly edentate catfish, belonging to a still-undescribed extra-large species.

Alternatively, as German cryptozoologist Markus Bühler has mentioned to me, South America's piraiba *Brachyplatystoma filamentosum*, a huge species of goliath catfish, has a deceptively shark-like body outline and is already known to grow up to 12 ft. long. So who knows, reports of "Gums" may have been based upon sightings of exceptionally large, geriatric piraibas.

THE WILD AMERICAN HOUND—A CANINE CONUNDRUM?

Here's an odd little conundrum for you to cogitate upon at your leisure, should you be so inclined. During an online surfing session in April 2012, I happened upon the curious illustration presented here.

Details concerning it are sparse in the extreme, but here is what I have been able to uncover so far. Measuring 12 inches by 8 inches, the image has a German title that translates as "wild American hound," and is a hand-colored copperplate engraving by Johann Daniel Meyer that appeared in his *Angenehmer und nützlicher Zeit-Vertreib mit Betrachtung curioser Vorstellungen allerhand kriechender, fliegender und schwimmender, auf dem Land und im Wasser sich befindender und nährender Thiere etc*, a three-volume wildlife tome published between 1748 and 1756 in Nuremberg, Germany.

As can be readily perceived from this engraving, however, whatever the beast depicted by it may be, it is certainly not a hound, nor, indeed, a canid, of any kind (wild and/or American notwithstanding!). So what is it?

When I first looked at it, I initially thought of the Virginia opossum *Didelphis virginiana*, because the engraved creature does bear a degree of overall resemblance to this largest and most famous of modern-day New World marsupials. I even found an online photo of the Virginia opossum that vaguely recalls it.

Even so, Meyer's mystery beast can be readily differentiated by its wholly brown coloration, in particular its dark face and its body's extremely short, uniformly brown fur—in stark contrast to the white face and the longer, shaggy, grey body fur of the Virginia opossum. Meyer's beast may have a bare tail, which, if so, likens it to the latter species, but, equally, it may simply have very short fur—the engraving does not make this clear.

Mid-1700s engraving depicting the "wild American hound" (public domain)

In addition to the Virginia opossum, I have also considered those uniformly brown-furred, Neotropical raccoon cousins known respectively as the kinkajou *Potos flavus* and the olingos (a quintet of *Bassaricyon* species). Again, these are somewhat similar to Meyer's beast superficially, but none is native to North America.

So unless the "American" in "wild American hound" was being used in its very broadest sense, i.e. appertaining to anywhere within the entire New World, rather than its much more common and more specific usage as a contraction of the United States of America, I have once again come to a halt in my search for this mystifying mammal's taxonomic identity.

Consequently, gentle readers, if you could offer any additional information or suggestions I'd very greatly welcome them—as indeed I also would for any of the other long-forgotten cryptids in this chapter.

Chapter 6:
SEA DRAGONS, FAIRY LOAVES, AND SERPENTS OF STONE— FABLES AND FOSSILS OF LYME REGIS

And by the shores of that unwholesome flood
There dwelt the mightiest of the scaly brood;
Huge dragons here indulged their murd'rous bent,
And loathsome lizards frolick'd in the scent
Of fens' most foul effluvium.

— Samuel Martins, "Tartarus"

In July 2010, I visited Lyme Regis on Dorset's south coast, a relatively small town but one that is to fossil enthusiasts what Hay-on-Wye on the Welsh-English border is to book lovers. Indeed, Lyme Regis is famous worldwide—and justifiably so—for the exceptional fossiliferous diversity and fecundity of its Mesozoic beaches and rocky cliffs, which have revealed untold paleontological treasures down through the ages—from fossilized skeletons of giant prehistoric marine reptiles such as ichthyosaurs and plesiosaurs, to a vast array of ancient invertebrates, including trilobites, ammonites, archaic crinoids or sea-lilies, and much else besides. Even so, the correct zoological identities of these remarkable relics from our planet's far-distant past have only been established by science during the last two centuries. Before then, as revealed here, the true nature of many of Lyme Regis's fossil fauna had inspired many fascinating legends and lore, as well as some very ingenious (albeit wholly inaccurate!) pre-scientific speculation.

CROCODILES, DRAGONS, AND MONSTERS GALORE!

Ichthyosaurs were specialized Mesozoic marine reptiles that due to convergent evolution were deceptively fish-like ("ichthyosaur" translates as "fish lizard") or even, in some cases, dolphin-like in superficial external appearance. They inhabited seas around the world from the mid-Triassic Period to the late Cretaceous Period (245 million to 90 million years ago), and many remarkably well-preserved specimens have emerged from the fossil-yielding rocks of Lyme Regis.

Indeed, the most famous single fossil specimen ever to be found here was an ichthyosaur. Moreover, in the eyes of many palaeontologists, it is still THE ichthyosaur—the 17-ft.-long skeleton that was discovered by local teenage fossil seeker Mary Anning in 1811. (Although, to be strictly accurate, its 3-ft. skull, the first part of the specimen to come to light, was actually discovered by Mary's brother, Joseph, after which Mary sought—and found—its skeleton.)

Mary went on to discover many other scientifically priceless specimens at Lyme Regis during the next three decades, including plesiosaurs and even an early pterosaur, *Dimorphodon*, which she sold to various scientific institutions as well as to wealthy natural history collectors. So eminent did Mary become in her own right that she was ultimately immortalized in a tortuous tongue-twister still familiar today: "She sells sea shells on the sea shore."

Today, paleontologists recognize that ichthyosaurs constitute a major reptilian lineage of their own, taxonomically discrete from all others. In earlier days, conversely, their spectacular remains incited much controversy and confusion among researchers attempting to categorize them. In the pre-scientific age, their fossilized relics were deemed to be the bones of antediluvian sea dragons that had survived the Great Flood and may still exist in the vast oceans even today, thus linking them directly with sightings of sea monsters and other maritime horrors of legend or cryptozoology (depending upon your personal opinion of such beasts).

In neighboring Somerset, a certain uncovered ichthyosaur skull was long deemed in local tradition to be that of a dragon called Blue Ben of Kilve, who served as the devil's own steed until it reputedly fell from a rocky causeway and drowned in the deep mud below. The skull can now be viewed in a nearby museum.

Mary Anning sold her ichthyosaur skeleton for the then-princely sum of £23 to Henry Host Henley of Sandringham, Norfolk, who was also Lord of the Manor of Colway, just behind Lyme Regis. He in turn deposited it in William Bullock's London Museum of Natural History, at Piccadilly. During the original excavations in November 1812 to remove its skeleton from the rocks, newspaper reports had referred to it as a petrified crocodile.

Although this may seem an unlikely misidentification to make, it must be remembered that, back in those days, no ichthyosaur remains presented for scientific scrutiny exhibited any impression of the prominent dorsal fin and caudal fin now known to have been sported by most ichthyosaur species—the first such fossils were not disinterred until the 1890s, in Germany. Without these finny accoutrements, therefore, to the casual untrained observer an ichthyosaur skeleton would indeed look somewhat crocodilian in basic form, especially with

19th-century engraving portraying a long-necked plesiosaur and a very sea-dragonesque ichthyosaur engaged in battle (public domain)

regard to its elongated jaws.

Conversely, when in 1814 the Mary Anning ichthyosaur was formally described by early paleontologist Everard Hume, he discounted a crocodilian identity for it but was just as erroneous in his own taxonomic judgment, because he considered it to be more closely allied to the fishes. Notwithstanding, when in May 1819 it was sold by Bullock's museum at auction, where it was purchased for £47 and 5 shillings by Charles Konig of the British Museum (it was Konig, incidentally, who had first coined the term "ichthyosaur," in 1813), it was listed as "a Crocodile in a Fossil State."

Although it is often claimed that the Mary Anning ichthyosaur was the first ever to be discovered, this is not true. Several notable, albeit incomplete, specimens had been documented prior to this (the earliest illustration of an ichthyosaur fossil—a vertebra from the Severn estuary—was published as long ago as 1699), and had been variously (but invariably) misidentified as the mortal remains of erstwhile whales, giant lizards, and even the (very) odd sea lion! The jaw of what was evidently a many-toothed ichthyosaur, obtained from the Dorset coast, was exhibited in London in 1783 at the Society for Promoting Natural History, but contemporary naturalists believed it to be from a fossil crocodile.

During the 1820s, scientific opinion swung away from fishes or crocodiles and towards whales as being the ichthyosaurs' closest living relatives. By the time that the finely preserved German fossils of the 1890s had been revealed, however, the ichthyosaurs' correct status as a separate, reptilian lineage had been established, and the dreams of biblical sea dragons were done with forever.

HEADLESS AND PETRIFIED—AND THAT'S JUST THE SERPENTS!

Had there really been sea dragons lurking offshore at Lyme Regis, they might well have become embroiled in some serious battles with the giant serpents also once deemed to exist here by local lore and fable. And constituting silent but very substantial evidence for these erstwhile claims are the countless specimens of what initially do resemble tightly-coiled stone serpents visible in prolific quantities embedded in the rocks on the beaches and emerging from the cliffs all about Lyme Regis and its environs. You can even walk directly over them, as I did, in the slabs of pavement constituting the Cobb—a curved mass of masonry sheltering the town's harbor. Further out is the famous Fossil Pavement composed of limestone, where untold numbers are encrusted. Many are tiny, but there are hundreds of much bigger ones too, some of which are almost the size of cartwheels—so, if uncoiled, these would have been serpents of truly prodigious size...but is this really what they were?

In bygone times, many people did genuinely consider them to be the petrified remains of the great sea serpents long believed to frequent the vast marine waterways of our planet. And, if so, perhaps these colossal beasts did engage in lethal confrontations with other maritime monsters after all, because one very curious characteristic shared by virtually every specimen of stony serpent, however great or small it may be, is that it is headless—a decapitated fossilized corpse devoid of skull, eyes, or mouth.

In truth, these cephalically challenged animals were not snakes at all, but ammonites—prehistoric cephalopod mollusks that inhabited Earth's ancient seas from the early Jurassic Period to the end of the Cretaceous Period (200 million to 65 million years ago). They were related to today's squids, cuttlefishes, and octopuses, but lived inside tightly whorled, ornately ridged shells (like the modern-day nautiluses, which are superficially ammonite-like in outward form but have much more fragile shells).

In life, their many-tentacled heads would have protruded out of their shells (which were usually, though not exclusively, planispiral in shape), but as the tissues constituting those structures are relatively soft, they are preserved far less successfully or abundantly than are shells, skeletons, and other harder materials during the fossilization process. Consequently, ammonite fossils exhibiting these details are much less readily encountered. Coupled with the ostensibly serpentine appearance of their shells, it can be readily understood how untrained pre-scientific observers may have assumed such fossils to be the remains of headless snakes.

Moreover, the very sizeable *Titanites* ammonites (often measuring 2 ft. or more in diameter) found in Dorset's Jurassic-dated Portland Stone greatly impressed the Isle of Portland quarrymen when encountering them in earlier ages, because they were convinced that these mega-fossils were the antiquated remains of gigantic eels. This explains why they are still colloquially referred to here as conger eels. Other giant ammonites of Dorset include *Arietites* and *Coroniceras*, found in Lyme Regis's Blue Lias limestone slabs on Monmouth Beach.

Discussing ammonite frauds and fables in my book *Karl Shuker's Alien Zoo: From the Pages of Fortean Times* (2010), I noted that the deceptive headlessness of these "stone serpents" inspired all manner of charming legends throughout Britain and elsewhere. These usually involved a local saint ridding the area of snakes by touching them, whereupon their heads immediately dropped off and their bodies then curled up and turned to stone. I also mentioned that when the sale of fossils became lucrative, some enterprising vendors would even skillfully carve snakeheads out

Arietites ammonite cast, Monmouth Beach (Karl Shuker)

of the stone matrix surrounding the open end of the ammonites' shells, thereby restoring these supposed serpents' lost heads! Clearly, it was not only the local saints who were capable of miracles!

THUNDER BULLETS, FAIRY LOAVES, AND DEVIL'S TOENAILS

Ammonites are not the only ancient cephalopod mollusks whose remains are liberally preserved within Lyme Regis's rocky repositories of fossils. Also represented here in prolific quantities, particularly amid the Black Ven Marls, are long, slender, bullet-shaped objects that have traditionally been referred to in southeast England as thunder bullets or thunderbolts, because they were believed to have been cast down from Heaven during thunderstorms. Elsewhere they have acquired even more memorable nicknames, such as scaur pencils in Whitby, northern England, jien-shih or sword stones in China, and vätteljus or gnomes' candles in Sweden.

In reality, they are none of these exotic items. Instead, they derive from prehistoric cuttlefish-related mollusks known as belemnites. Like ammonites, alongside which they died out at the end of the Cretaceous Period, belemnites sported tentacles (10) at the front of their body, encased inside a small-chambered cup-like shell or phragmocone but which was straight, rather than spiraled like the much larger shells of ammonites. Their body's rear portion was supported by a long bullet-shaped guard, believed to have served as a counterbalance device, and surviving during fossilization to yield the supposed thunder bullets encountered today.

Moving from gnomes' candles to fairy loaves, the latter are small heart-shaped fossils commonly found around Lyme Regis, especially in 100-million-year-old Cretaceous Upper Greensand flints lying on the beaches. It was once believed that they were not only loaves created by fairies but also imbued with fairy magic themselves, for according to traditional southern England folklore a family whose house contained one or more of these curious little objects would never be without real bread, even in times of shortages or famine, and their milk would never turn sour either. Eventually, however, such quaint legends were brushed aside by the rather more prosaic scientific reality that these were in fact the fossilized shells or tests of *Micraster*, a sea urchin.

An interesting example of Dorset folklore neatly linking belemnites and fossil sea urchins was recorded in Vol. 2 of English antiquarian John

Sea Dragons, Fairy Loaves, and Serpents of Stone

Brand's *Observations on the Popular Antiquities of Great Britain* (1777):

> In Dorset the Pixy-lore still lingers. The being is called Pexy and Colepexy. The fossil belemnites are named Colepexies-fingers; and the fossil echini, Colepexies-heads. The children, when naughty, are also threatened with the Pexy, who is supposed to haunt woods and coppices.

As revealed by its alternative name of Colt-Pexy, this entity often assumes the form of a mischievous pony. Of course, in such a guise it would sport hooves, not fingers, but folklore has never been a great respecter of anatomical or zoological niceties!

Equally, devil's toenails, these very distinctive-looking objects commonly encountered in Lower Lias (Jurassic) rocks up and down Great Britain, including the Blue Lias of Lyme Regis, are actually the large, fossilized incurved left valves or shells of specimens belonging to a species of prehistoric marine bivalve mollusk known as *Gryphaea arcuata*, which was ancestral to today's oysters. The right valves of *Gryphaea* specimens were much smaller, flat, and lid-like, thus attracting less attention from casual observers.

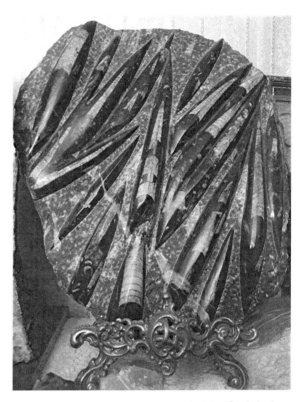

A slab of polished fossil belemnites (Karl Shuker)

Today, the East Devon and Dorset coastline incorporating Lyme Regis's fossil-rich repositories is known as the Jurassic Coast and has very deservedly been listed as a natural World Heritage Site (England's first), with its spectacular wealth of paleontological treasures internationally renowned and extensively documented in the palaeo-literature. Happily, however, their scientific eminence has not entirely banished their associated folkloric traditions from the modern world, which persist just as tenaciously as the fossils themselves. And rightly so—after all, a planet without sea dragons, fairy loaves, thunder bullets, and stony serpents sans heads would certainly be an infinitely drabber, sadder one!

Chapter 7:
GIANT LEMURS AND DWARF HIPPOS—ON THE TRACK OF MADAGASCAN MYSTERY BEASTS

Madagascar is the land of sub-fossil creatures. No other place in the world gives so strong an impression of having just emerged from Prehistory. The excavated remains of huge animals, whether they be birds, reptiles or mammals, are sometimes so fresh that one cannot help wondering whether the species they come from are really extinct. The native traditions often make it perfectly clear that they were known by man until quite recently.

— Bernard Heuvelmans, *On the Track of Unknown Animals*

The fourth largest island in the world, Madagascar is a veritable miniature continent, brimming with extraordinary animals found nowhere else in the world. These include such exotic avifauna as vangas, mesites, and asities; remarkable reptiles like boas that are more closely related to counterparts in South America than to any Old World snakes, and the world's greatest diversity of chameleons; plus, most famous of all, its wonderful assemblage of lemurs and exotic striped civets. Yet, as if all of these were not exciting enough, only a few centuries ago in some cases this zoological treasure trove of endemics also boasted some truly astonishing animals—such as an impressive number of giant lemurs far bigger than anything known to exist today, three different species of pocket-sized hippopotamus, and a massive flightless bird of such gargantuan proportions that it was aptly dubbed the elephant bird. Today, of course, these creatures are long gone—or are they? Read this investigation of Madagascar's mystery beasts—the most extensive ever published—and decide for yourself.

TRATRATRATRA GOODBYE...?

In 1658, after having resided in Madagascar for a number of years as its governor, French explorer Admiral Étienne de Flacourt published a major tome, *Histoire de la Grande Îsle de Madagascar*, which documented his experiences and discoveries here. It contains a wealth of zoological information, some of which is also of cryptozoological interest, including the following passage:

> Trétrétrétré [or tratratratra in English] is an animal as big as a two-year-old calf, with a round head and a man's face; the forefeet are like an ape's, and so are the hindfeet. It has frizzy hair, a short tail and ears like a man's...One has been seen near the Lipomani lagoon in the

neighbourhood of which it lives. It is a very solitary animal, the people of the country are very frightened of it and run from it as it does from them.

At the time, European naturalists dismissed this report as native folklore. Also discounted were the description and accompanying sketch of what may have been the same, or at least a very similar, unidentified species, but in this instance called the thanacth, documented by André de Thévet in his *Cosmographie Universelle* (1575). The creature in question had been brought to him, while based on the Red Sea, as a curiosity by natives from an unspecified eastern land that some authorities now believe may have been Madagascar.

A few centuries later, however, paleontologists in Madagascar began unearthing fossilized remains of enormous lemurs, which, when dated, proved in some cases to be from creatures that had only died out a few hundred years earlier. Moreover, reconstructions of the likely appearance in life of certain of these animals seemed more than a little reminiscent of the tratratratra.

Sketch of the thanacth in André de Thévet's *Cosmographie Universelle* (public domain)

Consequently, in his seminal cryptozoological book *On the Track of Unknown Animals* (1958), zoologist Bernard Heuvelmans boldly proposed that this mystery beast may indeed have been a surviving representative of the sloth lemur *Palaeopropithecus ingens*, a giant lemur shown from radiocarbon dating of subfossil remains to have existed until at least the 1500s. As big as a chimpanzee, somewhat sloth-like when in trees but probably at least partly terrestrial due to its large size and weight, *Palaeopropithecus* would have appeared spectacular and somewhat awe-inspiring to the native people—hence, if indeed synonymous with the tratratratra, their fear of it.

However, it is now known that *Palaeopropithecus* had a very pronounced snout, which contrasts with the man-like (and thus presumably flattened) face described by de Flacourt for the tratratratra. Conversely, a second identity offered for it by Heuvelmans, the large extinct lemur *Hadropithecus*, did have a relatively flattened, ape-like or humanoid face (as opposed to the long-muzzled, canine faces of most known modern-day lemurs), and is known to have existed until around a thousand years ago.

It has lately been confirmed that the giant lemurs were hunted—quite probably into extinction—by humans (who first reached Madagascar around 2,000 years Before Present). In April 2002, at a meeting of the American Association of Physical Anthropology, a team of scientists from the University of Massachusetts and Oxford's Natural History Museum revealed that some remains of *Palaeopropithecus ingens* and *Megaladapis* (another extinct giant lemur, see below) from Taolambiby (a subfossil site in southwestern Madagascar), which were originally collected back in 1911 but only lately studied, showed classic signs of butchering. The characteristics of the tool-induced bone alterations (sharp cuts near joints, spiral fractures, and percussion striae) suggested dismembering, skinning, and filleting.

Reconstruction of the sloth lemur *Palaeopropithecus ingens* (Markus Bühler)

Moreover, the giant lemurs also suffered from habitat destruction via extensive deforestation. Nevertheless, there are still areas of dense, remote Madagascan forest little visited by humankind even today, where reports of bizarre beasts continue to emerge from time to time. One of the most pertinent of these reports relative to the tratratratra controversy was published in Jane Wilson's fascinating book *Lemurs of the Lost World* (1990). After mentioning de Flacourt's account of this cryptid, she notes:

> Although this description may be distorted, it is the last accepted sighting of the now extinct giant lemurs. A few may have survived until the 1930s, however, when a French forester came face to face with an animal sitting four feet high and described it as being unlike other lemurs he had seen. It did not have a muzzle but was like a gorilla with "the face of one of my ancestors."

Or the face of the last tratratratra, perhaps?

Equally fascinating is the following recent account of an alleged tratratratra, from Arnošt Vašíček's book *Planeta Záhad: Tajemná*

Minulost (1998), which has been kindly translated from Czech into English for me by Czech friend and crypto-colleague Miroslav ("Mirek") Fišmeister:

In 1991, two men from the Betsimisaraka tribe stayed overnight in the forest by the foot of the Tsaratanana mountain range. They made their camp under a rock overhang, lit a fire and started to eat what they brought with them. Suddenly they smelled an intense animal odour. They could not see anything in the darkness. Only the sound of breaking sticks proved that something heavy was approaching them carefully. They heard excited breathing. Then something clicked its tongue strangely. After a long time, they could recognise a large body with broad shoulders and rather small, pointed head, where the light was turning into a shadow. The giant was over 6.5 ft tall. It had long, strong arms. It was gangling and did not move, as it was unsure whether to come closer or run away. The fire did not seem to frighten it. It was rather the presence of people which made it wonder-struck and unsure. The Betsimisarakas could not see its face, but they had the feeling, as they admitted later, that they were being observed by the tratratratra with curiosity. How long this encounter was, they could not estimate. They only knew that at last they overcame their fear and threw their food towards the tratratratra. The fire began to fade. They reached for the branches they had prepared before, but before the fire had started to glow again and the flames lit the night, the creature left the glade. They did not dare to rise and sat there numbly until the morning. The food which they tried to redeem themselves with was gone. However, it was uncertain whether it was taken away by the tratratratra or eaten during the night by animals.

One noon, a tratratratra surprised a group of girls with its roar by the Masora river. It was probably trying to cross the river or simply take a bath, but the strong stream pulled it to the deep water. Frightened, it tried not to sink and reach the bank. It floated for several tens of metres [yards], before it managed to grab a large boulder in front of the arriving group. The girl saw its face with broad flat nose and protracted jaws. The deep eyes under big brow ridges were full of fear. The tratratratra was leaning against the boulder and tried to grab a thick branch of a tree bent over the river. It was wailing desperately, as if it was asking the girls for help or tried to call its fellows. After it did not manage to grab the branch, it started to try to climb up onto the rock. The girls panicked and ran away. The local shaman and several men from the village immediately organised a rescue party, but when they came to the river, there was no-one on the rock.

Although unlikely, it is not impossible that a very small, relict population of at least one species of giant lemur does still persist in Madagascar—highly elusive, nocturnal, and normally avoiding humans whenever possible.

ON THE TRACK OF THE TOKANDIA

In earlier works, the tratratratra was often synonymized with another giant Madagascan lemur, the so-called koala lemur *Megaladapis edwardsi*. Like *Palaeopropithecus*, this species is now known from dated subfossil remains to have still existed as recently as 1500 AD, and may well have met its demise at the hands—and weapons—of humans.

It derives its name from the outward similarity of its general body form to that of Australia's familiar koala *Phascolarctos cinereus*—but a koala on a gigantic scale, as this tree-dwelling lemur sported a skull the size of a gorilla's and weighed a massive 165 lb or so. It had proportionately long forearms, extraordinarily cow-like jaws, a very elongate face, huge grasping hands and feet, and quite possibly a short tapir-like nasal trunk. Although clearly adapted for life in the trees, its huge size indicated that this monstrous lemur might well have spent quite a lot of time on the ground.

The extended shape of its face and heavy bovine jaws clearly argue against *Megaladapis* being one and the same as the tratratratra, and in later publications this identification has indeed been discounted in favor of *Palaeopropithecus* or *Hadropithecus*, as already discussed here. However, there is another Madagascan mystery beast that *Megaladapis* does compare well with—a huge, largely terrestrial mammal called the tokandia.

Reconstruction of the koala lemur *Megaladapis edwardsi* (Markus Bühler)

Just like modern-day sifakas and certain other lemurs, the tokandia is said to move on the ground via a series of bounds or leaps, but also jumps into trees, where it spends time, too. Moreover, unlike the tratratratra, the face of the tokandia is claimed by the locals not to be man-like, but its cries are allegedly very like those of humans. Accordingly, Heuvelmans and other cryptozoologists have identified the tokandia with *Megaladapis*. Whether it still exists today, conversely, is another matter entirely, as there do not seem to be any modern-day accounts of the tokandia. Also, even in the least-accessible surviving forests of Madagascar, a koala-shaped lemur the size of a bear would surely be somewhat difficult to overlook.

NO KIDDING, IT WAS A KIDOKY

Having said that, there is tantalizing evidence that some other form of very large, still-undiscovered species of lemur does still exist in this insular mini-continent. During late July and early August 1995, Fordham University biologist David A. Burney and Madagascan archaeologist Ramilisonina conducted ethnographical research at three remote southwestern Madagascan coastal villages, in particular Belo-sur-mer. Here, interviewing the local people, they collected testimony from eyewitnesses describing three different mysterious beasts, two of which may well be species still unknown to science.

One of these creatures was referred to as the kidoky, and according to consistent local descriptions (obtained by interviewing the eyewitnesses completely independently of one another) it apparently resembles those relatively large, principally tree-dwelling lemurs known as sifakas. However, in terms of overall size, the kidoky was said to be much larger.

Just such an animal was reputedly sighted as recently as 1952, by an educated villager named Jean Noelson Pascou. When interviewed by the two scientists, Pascou was adamant about what he had seen, and stated that the kidoky had dark fur, but with a white spot below its mouth and another one upon its brow.

When on the ground, sifakas move via a very characteristic series of sideways bipedal bounds and if threatened will flee up into the trees. In contrast, the kidoky allegedly flees by running away in a series of short, forward leaps reminiscent of a baboon's mode of locomotion, and usually remains on the ground, rather than taking to the trees. Its face is quite round and man-like, but its loud whooping call is more similar to that of the decidedly dog-headed indri *Indri indri* (currently the largest known species of living lemur).

As noted by the two scientists, in terms of overall morphology and lifestyle the kidoky brings to mind two officially extinct genera of giant lemur—*Archaeolemur* and *Hadropithecus*. Both exhibited terrestrial adaptations and were the closest equivalents in lemur terms to the baboons. Like other giant lemurs, however, they are assumed to have died out several centuries ago. But if the testimony of Pascou and other Belo-sur-mer villagers are to be believed, this assumption may well be premature.

THE HABÉBY—A LEMUR IN SHEEP'S CLOTHING?

One of the most mystifying of all Madagascan cryptids is the habéby. Also called the fotsiaondré, it is likened both in size and in overall appearance to a large white sheep, with long furry ears, and an elongate muzzle, whose coat is dappled with brown or black, and whose feet are reputedly cloven. The Betsileo tribe avers that it inhabits the wastelands of the Isalo range, and that it can sometimes be seen on moonlit nights. For—bizarrely for any sheep—the habéby is claimed to be strictly nocturnal, which presumably explains why it is also said to have very large staring eyes, another most unsheep-like characteristic. Equally odd, if it is indeed a sheep, is that there has never been any report or any tradition of horned habébys, i.e. habéby rams.

Faced with these disconcerting inconsistencies with any typical ovine identity, it is little wonder that zoologists have considered it much more feasible that the habéby is (or was) an elusive species of very large terrestrial lemur, which would explain its nighttime activity and associated large eyes. Of course, lemurs are not cloven-footed, but perhaps a predominantly terrestrial, giant form may have evolved superficially hoof-like claws to assist its locomotion on the ground.

German cryptozoologist Markus Bühler has examined the habéby case in detail, and in January 2011 he offered me the following interesting thoughts on it:

> It is extremely probable that the habéby is based on *Megaladapis*, not only the description of the eyes, the fur and the nocturnal habit indicates a lemur, but especially the size and comparison with a sheep. The largest species *Megaladapis edwardsi* perfectly fits the size of a sheep (the old weights of more than 100 kg [220 lb] are very improbable), and most of all the shape of the head (and...the lack of a prominent tail). The skull is really extremely elongated for a lemur, and more than anything else it feels like those of an ungulate. I have already handled skulls of chimpanzees, gorillas, baboons and some other monkeys and apes, as well as those of some other larger mammals including lions and bears. Last year I dissected a sheep head, and it was really fascinating, and helped me a lot in the understanding of animal anatomy. The

skull has of course differences to those of *Megaladapis*, but also a lot of similarities. I think given the fact that there aren't that many alternative large animals in Madagascar, *Megaladapis* is really by far the best explanation for "the white sheep." Heuvelmans already proposed a lemur as identity, but didn't suggest any species...I decided to give the head of the *Megaladapis* model [one of several cryptozoological models that Markus has produced] colours which fit the description of the habéby, with a very bright (whitish-grey) fur and only some minor dark markings...Of course the large canines do not look very sheep-like, but in a closed mouth, especially in the dark, they would be hardly visible.

MAKING AN ASS OF THE MANGARSAHOC

Equally unexpected in Madagascar are reports of a mysterious white ass with huge ears that almost cover its face. This unlikely-sounding beast is referred to as the mangarsahoc, and, like the tratratratra, was briefly documented by Admiral de Flacourt in his great tome:

> Mangarsahoc is a very large beast, which has a round foot like a horse's and very long ears; when it comes down a mountain it can hardly see before it, because its ears hide its eyes; it makes a loud cry in the manner of an ass. I think it may be a wild ass.

In 1770, this strange creature was also reported by the Comte de Modave, who claimed that it lived some 10 leagues from Fort Dauphin, and that the local tribespeople were terrified of it:

> [Yet] at bottom it is but a wild ass; many are found in this part of the island, but you must look for them in the woods, for they never leave these lonely places and are hard to approach.

Intriguingly, tracks of hooves said to be from the mangarsahoc have actually been found, and several sightings have been reported in the Ankaizinana forests as well as in the Bealanana and Manirenjy districts, according to Heuvelmans. Yet no specimen has ever been obtained, and, as with the habéby, Heuvelmans was more inclined to deem this evanescent creature a giant lemur than a hoofed, ungulate mammal. Tellingly, it has a vile reputation among the native people, who firmly believe that the mere sight of it will bring bad luck. This seems an unusual superstition to become attached to a wild horse, yet is of the very same kind often associated with some of the more striking, nocturnal lemurs—in particular, the rather eerie-looking (yet wholly harmless, inoffensive) aye-aye *Daubentonia madagascariensis*.

So could the mangarsahoc also be an undiscovered species of very large, terrestrial, pseudo-hoofed lemur? Or might it instead be a non-existent composite beast, engendered by confusion between Madagascar's extra-large lemurine cryptids and a very different mystery beast, the tongue-twistingly-named kilopilopitsofy?

WHAT'S IN A (VERY LONG) NAME? THE KILOPILOPITSOFY

Also known as the tsomgomby, the kilopilopitsofy was the second unidentified beast spoken of by the Belo-sur-mer villagers to scientists David A. Burney and Ramilisonina during their ethnographical researches.

As with the kidoky, they received consistent reports of this other mystery animal from several different eyewitnesses, including, once again, Jean Noelson Pascou, who was able to describe the kilopilopitsofy in considerable detail, following his own sightings of it, one of which was made as recently as 1976. According to Pascou, it is cow-sized but hornless; has very dark skin, pink coloration around its eyes and mouth, fairly large floppy ears, big teeth, and large flat feet; and is nocturnal, and escapes from danger by running into the water.

When shown photos of various animals, another kilopilopitsofy eyewitness selected a hippopotamus as bearing the closest resemblance to the kilopilopitsofy. Moreover, when asked to describe the sounds that this mystery beast makes, Pascou, who was well known in the village as a skilled imitator of local animal noises, gave voice to a series of deep, drawn-out grunts, which the startled scientists realized were very similar to the sounds made by the common hippopotamus *Hippopotamus amphibius* of the African mainland. Yet no species of hippopotamus is supposed to exist on Madagascar—or not any longer, that is.

In fact, living alongside the giant lemurs on Madagascar were once no less than three different species of endemic Malagasy hippopotamus. These were: the lesser Malagasy hippo *Hippopotamus laloumena* (a small relative of the common African mainland species); the Malagasy dwarf hippo *H. lemerlei* (an even smaller, dwarf species); and the Malagasy pygmy hippo *Hexaprotodon madagascariensis* (a distinctive pint-sized species more closely related to the African mainland pygmy hippo *Hexaprotodon* [=*Choeropsis*] *liberiensis*). Just like the giant lemurs finally became extinct, however, so too did all three Madagascan hippo species, once again having been hunted and eaten by humans, but with subfossil evidence confirming that at least one species, the Malagasy dwarf hippo *H. lemerlei*, was still alive as recently as 1000 BP.

Furthermore, judging from the testimony of Pascou and others with regard to the mystifying kilopilopitsofy, zoologists as well as cryptozoologists now consider it plausible that at least one of Madagascar's diminutive trio of hippos actually persisted into much more recent times, and perhaps even into the present day. As for this mystery beast's floppy ears, they deem it possible that these were actually loose jowls and cheeks, misidentified as ears when seen fleetingly at night. Alternatively, it may be that they really are ears, larger than those of typical hippos, which have evolved to help dissipate heat if, as seems to be true, the dwarf hippo in particular was (or is?) more terrestrial than typical hippos.

In 1876, the skin of an alleged tsomgomby was shown to Westerner Josef-Peter Audebert, who likened it to that of an antelope (no species of which exists on Madagascar), and stated that it had supposedly come from the south of the island. If only this skin had been preserved—DNA analyses could have unmasked its owner's identity and thereby resolved the longstanding mystery of this very perplexing Madagascan cryptid.

KALANORO—MADAGASCAR'S LITTLEFOOT

Perhaps it is only fitting that on such a cryptozoologically paradoxical island as Madagascar, where there are reputed sightings of giant lemurs and dwarf hippos, there should also be observations not of a bigfoot-type man-beast comparable to those reported from many other regions of the world, but rather a littlefoot—an elusive hairy ape-man, yet of only very short height. Known as the kalanoro (aka kolanaro), many accounts of it exist, including the following detailed example, published in 1886 by G. Herbert Smith within the *Antananarivo Annual*:

> We next come to the forest, and from there we get endless stories of the Kalanoro, a sort of wild-man-of-the-woods, represented as very short of stature, covered with hair, with flowing beard, in the case of the male, and with an amiable weakness for the warmth of a fire. An

eye-witness related that once, when spending a night in the heart of the forest, he lay awake watching the fire, which had died down to red embers, when suddenly he became aware of a figure answering to the above description warming himself at the fire, and apparently enjoying it immensely. According to his story, he put a summary end to the gentleman's enjoyment by stealing down his hand, grasping a stick, and sending a shower of red-hot embers on to his unclothed visitor, who immediately, and most naturally, fled with a shriek. Another tells how, on a similar occasion, the male appeared first, and after inspecting the premises and finding, as well as a fire, some rice left in the pot, summoned his better half; the pair squatted in front of the fire and—touching picture of conjugal affection—proceeded to feed one another!

One must confess that the creature described looks suspiciously like one of the larger sorts of lemur; but in a village near Mahanoro [in eastern Madagascar], and on the verge of the forest, the inhabitants say that very frequently these wild people come foraging in their houses for remnants of food, and may be heard calling to one another in the street.

In the 1930s, French palaeontologist Charles Lamberton speculated that perhaps the kalanoro was based upon folk memories of the last *Hadropithecus*, the large extinct lemur mentioned earlier in this chapter that had a remarkably human profile. However, I am not convinced by a lemur identity for this cryptid, and I certainly disagree with Smith that the kalanoro he described in his reports resembles a large lemur. On the contrary, it seems much more humanoid than lemurine, and nowhere was there any mention of a tail. Yet with the exception of the largest, officially living lemur, the near-tailless indri, lemurs generally have very lengthy, noticeable tails.

So unless there is a totally unknown, dramatically different species of lemur out there whose evolution has yielded a veritable human counterpart, tailless and bipedal, it seems much more likely that if the kalanoro is more than just a Madagascan counterpart of the Western world's Little People or fairy folk (or even mermaid—some kalanoro accounts claim that it is amphibious and female!), it may well be a primitive form of human.

Certainly, it bears more than a passing resemblance to reports of Sumatra's elusive orang pendek or "short man," which in turn has been associated lately with the startling discovery on the nearby Indonesian island of Flores of a dwarf fossil species of human, dubbed Flores man or the hobbit *Homo floresiensis*. Could the kalanoro be something similar? Tragically, we may never know, for although it was clearly once a well-known entity on Madagascar, the days when it would visit people's fires and forage in their houses for food seem long gone—just like, apparently, the kalanoro itself. Unless, that is, it does still persist in Madagascar, but is known by a different name—the kotoko?

INTRODUCING THE KOTOKO

As far as I can tell, this Madagascan cryptid has never previously been documented in any cryptozoology book. Consequently, I am greatly indebted to correspondent James Skinner for so kindly permitting me to include here the following information concerning it that he recently made available to me.

James is the son of veteran British fortean researcher Bob Skinner (who also generously assisted me by providing kotoko information passed on by James), and he is currently living in Madagascar, where he works as a charity volunteer. During his spare time, he has been investigating native reports of a small hairy humanoid entity known to them as the kotoko. The first person to tell him about this being was a man named Aime, from the Mahatalaky Commune in the Anosy Region of southeastern Madagascar, who also prepared the sketch on p. 83 of the kotoko's alleged appearance.

In September 2012, James interviewed Aime on video concerning the kotoko, then posted it to *YouTube* on November 18, 2012 (http://www.youtube.com/watch?v=VCh9Yf4E81g&).

In this videoed interview, Aime stated that the kotoko is humanoid in form but hairy, with a big chest and small feet, and that his grandparents and other elderly inhabitants of his village have seen males and females. He also claimed that there have been sightings as recently as just a few years ago. The kotoko steals food from villages at night, and fish from fishermen, especially in more remote regions, but will soon vanish if it hears unexpected noises, as it is very swift moving.

Intriguingly, there is a tradition that the kotoko, which is very strong too, will wrestle with villagers who are good wrestlers. Aime also claimed that it is even possible to make friends with the kotoko. If rice is left outside villagers' homes for it on a regular basis, the kotoko will reciprocate by leaving for the villagers some honey or dead tenrecs (small insectivorous Madagascan mammals popularly killed and eaten here for their meat).

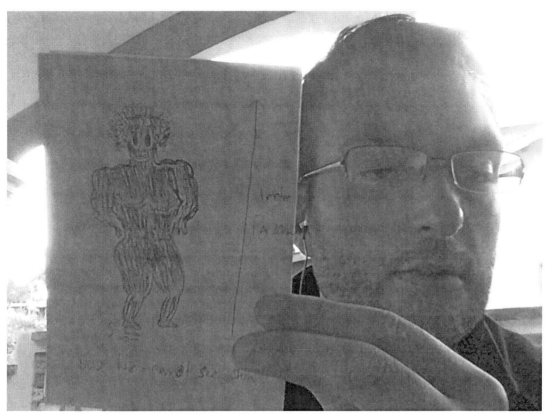

Aime's sketch of a kotoko, held by James Skinner (Aime and James & Bob Skinner)

A couple of months after videotaping the interview with Aime, James spoke to an old man living in a far-flung village regarding the kotoko, and here is the full report of this incident that James subsequently prepared:

Last Friday (9th Nov) I headed to the bush for a project celebration organised by some of my colleagues in a really remote village called Beandry, a 15 minute walk from the nearest road (including a risky river crossing!).

Aime the colleague who was the first to tell me the stories of the Kotoko several months ago was at the event. On Friday he introduced me to an old man who lives locally and translated my questions as we spoke.

The man seemed to have a lot of stories about friends and relatives who had experienced this phenomenon. He said the Kotoko are known to live on and/or near the hills that we could see in the distance from where we were chatting in the village. He described the Kotoko as being short and completely covered with hair as I had heard before, although there was no mention of their eyes lying vertically and he said their feet face forward unlike some of the reports shared by Aime previously.

He gave an interesting explanation for the existence of these mysterious little people claiming that Kotoko came into being during the French colonial presence (1900-1960). This theory (very similar to one told to me by Aime before) claims that some villagers escaped to the hilly forests to escape paying the taxes imposed by the French and that over time they had become wild and evolved into the Kotoko.

He gave several interesting accounts of encounters that he had heard of:

- About 15 years ago three of his relatives caught one female Kotoko and held it for some time.

- One man was walking near a waterfall when he discovered a Kotoko who had his arm stuck under a rock. He helped remove the rock and free the Kotoko, who was very grateful. The Kotoko then started bringing gifts of honey etc. for the man and his family. One day the man trapped the Kotoko in his house so he could show him to his fellow villagers. This angered the Kotoko who then attacked the man with his long sharp nails.

- He confirmed that the Kotoko have been known to wrestle with the wrestling champion in any village they come across.

Aime suggested that if we were to go to the hills we could attract the Kotoko by putting out some food and that if this was spiked with alcohol it would send the Kotoko to sleep and we would have a good chance of getting a look at them up close!

In an email to me of December 9, 2012, responding to an earlier one of mine in which I provided him with information concerning the kalanoro, James offered the following insightful thoughts and opinions regarding the kotoko and kalanoro:

Very interesting to read the info you have about the Kolanaro so far, it definitely sounds very similar to the Kotoko. I'd seen references to Kolanaro online before, and checked with Aime (my main contact) about this when I filmed a second interview with him last week, he had not heard the name Kolanaro and suggested that it may be a different name for the same creature.

In the interview he explained a bit more about a couple of cases where people claim to have caught the Kotoko.

I also asked him to explain more about the Kotoko's appearance and he explained that although some people have said they have backward-facing feet and eyes that run vertically unlike ours others have claimed that the feet and eyes were more conventionally human-like.

He also explained about some other types of wild men that are reported to exist here (I'll revisit the interview with him to get the names of these).

I would be more than happy for you to include the accounts of the Kotoko also, this corner of Madagascar is not as well publicised as others so it's great if the local folklore can be reported on (I wouldn't be surprised if other regions have different names too).

Man-beast researchers will recognize the reference to backward-pointing feet, because this is a motif in such reports that crops up time and again, in accounts from all over the globe. As noted here by James, however, not everyone believes that this feature is genuine—it seems to owe more to traditional folklore than to zoological fact. And if, as James believes, the kalanoro and kotoko are one and the same entity, this is exciting news indeed, as it means that although this cryptid could well be extinct in the regions where it is dubbed the kalanoro, in those where it is known as the kotoko it may still survive today. James plans to continue his researches into this fascinating, ostensibly resurrected cryptozoological entity. So perhaps even more information will be forthcoming soon.

FUSSING OVER A FOSSA

Known zoologically as *Cryptoprocta ferox*, the fossa is Madagascar's largest native mammalian carnivore and superficially resembles a small long-tailed puma, but is more closely related to the civets and genets (viverrids). It has a total length of 4.5-5.5 ft., but there was once an even bigger species, the giant fossa *C. spelea*, which measured a formidable 6.5 ft. long. Although officially believed to have died out at least several centuries ago, reports of a very large, fierce cat-like beast stalking this island's most inaccessible forests and known locally as the antamba have led to speculation that perhaps the giant fossa may still exist. As might be expected, the antamba did not escape the attention of the indefatigable Admiral de Flacourt, who documented it as follows:

The Madagascar antamba is an animal as large as a dog, with a round head, and, according to the relations of the Negroes, resembling the leopard. It devours both men and cattle, and is only found in the most desert[ed] parts of the island.

In November 1999, fossa expert Luke Dollar, intrigued by persistent rumors concerning the antamba, trekked in search of it through northeastern Madagascar's Zahamena National Park, also known as the Impenetrable Forest, which is said to harbor this elusive beast. Unfortunately, he did not sight it but has not given up hope that it does exist, as most of this uninviting region has never been scientifically explored.

Also worth noting is that, although once again still-unconfirmed by science, there have been a number of reports of melanistic (all-black) fossas in certain eastern rainforest areas of Madagascar.

TOMPONDRANO—MADAGASCAR'S SCALY LORD-OF-THE-SEA

It is, perhaps, inevitable that Madagascar's answer to the sea serpent is very different from those reported elsewhere in the world. Known as the tompondrano or Lord-of-the-Sea, it was described as follows by the Bezo tribespeople to the Paris Museum laboratory's then assistant director, Georges Petit, who documented it in his book *L'Industrie des Pêches à Madagascar* (1930):

> The Lord-of-the-Sea appears rarely. But he shows himself, whenever the time may be, by always moving against the wind. He is 70 to 80 feet long, and his wide flat body is covered with hard plates, rather like the bony armour on the back of a crocodile, but bigger. The tail is like a shrimp's tail with its terminal flap. The mouth is ventral, the animal must turn on its back to attack. A sort of hood which the animal may raise or lower at will protects its eyes which look forwards but are placed well to the side. The head is luminous and shines light as it comes to the surface. It moves in vertical undulations.
>
> Some Malagasies say the animal has no legs. Others say it has front flippers like a whale's. Finally the body is striped in a longitudinal direction, with stripes of different colours, white, red, green or darker. It has no smell. The most frequent appearances occur in the area of Ankilibé, at Nosy Vé, and Lanivato (Tuléar province).
>
> To avoid the dangers which the tompondrano causes, an axe and a

silver ring are hung on the boat's bows.

In his book *In the Wake of the Sea-Serpents* (1968), Bernard Heuvelmans postulated that the tompondrano was a surviving species of armoured archaeocete (archaeocetes were primitive and sometimes very elongate species of whale). But it is now known that archaeocetes were not armored (this former, mistaken belief derived from some fossilized archaeocete remains that had been found in association with scales from a completely different animal). Instead, as I proposed in my own book, *In Search of Prehistoric Survivors* (1995), it is more plausible that scaly sea serpents like the tompondrano and also the Vietnamese con rit or sea millipede may well be giant shrimp-like crustaceans, whose huge size would be readily buoyed by their seawater medium.

Incidentally, despite statements made on at least one cryptozoological website, I have never claimed that this cryptid might be a marine myriapod (centipede or millipede); this error on the website's part may have derived from the fact that commonly-used (albeit zoologically-misleading) English names for the tompondrano include sea centipede and sea millipede. Nor have I specified which type of crustacean I consider it may represent (this same website erroneously stated that I believed it to be an isopod crustacean); without a body to examine, any such line of speculation is worthless.

UNMASKING MADAGASCAR'S DOG-HEADED MEN

Confronted by so many mystifying creatures reported from Madagascar, it is good to know that at least one of them has been formally identified and admitted into the catalogue of recognized, accepted zoological species.

Centuries ago, Ctesias, Marco Polo, and a number of other famous travelers claimed that certain islands to the east of Africa harbored a grotesque race of semi-humans, which had the bodies of men but the heads of dogs. These weird entities were thus referred to as cynocephali—"dog-heads"—but for a long time they were dismissed by naturalists as being wholly mythical.

When scientists began exploring Madagascar, however, they discovered to their considerable surprise that this great island's cynocephali really did exist. They proved to be a very large species of almost tailless lemur, the largest lemur in fact known to exist today, nearly 4 ft. tall, called the indri or babakoto *Indri indri*, and revered by certain native tribes who deem it to be a former human being. Although its body is quite humanoid, especially when it is seen sitting upright high in a tree, its head is extraordinarily canine in shape—a veritable cynocephalus or dog-head, a legend come to life.

THE VOROMPATRA—FROM ELEPHANT SNATCHER TO ELEPHANT BIRD?

In the Arabian Nights fairy tales, Sinbad the sailor visited an island in the East frequented by a monstrous eagle-like bird—a bird so enormous and powerful that it could snatch up a full-grown elephant with ease in its mighty talons and bear the hapless beast aloft before dropping it to the ground to kill it, after which it would carry the elephant's remains to its nest and feed its flesh to its ravenous young. This winged wonder was known as the roc or rukh. Yet although widely deemed to be a wholly fabulous, totally imaginary entity, several real-life medieval travelers claimed to have heard tell or even to have seen physical evidence for the existence of just such a creature.

Raffia leaf, which might once have been claimed by wily merchants to be a roc feather (Karl Shuker)

Perhaps the most famous of these personages was none other than Marco Polo, who affirmed that, according to his Asian host the Mongol emperor Kublai Khan, the roc lived on the island of Madagascar or isles close to it. He also stated that the emperor had shown him an immensely long roc feather and two gigantic roc eggs, all of which were owned by the aforementioned emperor.

It is highly likely that the so-called roc feather was actually one of the huge leaves of the raffia palm tree *Raphia regalis*, which are deceptively plume-like—as seen in the photograph above of one such leaf, from my personal collection of zoological curios. In medieval times, these leaves were often mistaken for (or passed off by wily merchants as) giant feathers, especially by (or to) gullible Western travelers and pilgrims to the Middle East, Africa, and the Orient. The giant eggs, conversely, are much more plausible, because during the early 1800s many exceptionally large egg fragments, and even some complete eggs, again of truly colossal size, were indeed discovered by scientists on—where else?—the island of Madagascar.

Until then, the largest bird eggs known were those of the ostrich *Struthio camelus*, but the gargantuan specimens found on Madagascar were so immense, with a capacity of up to a gallon, that they could hold no less than six ostrich eggs! Moreover, during the 1860s some bird bones disinterred from marshes on this same island were so massive that the ostensibly extinct species from which they had originated soon became known as the great elephant bird, and was scientifically dubbed *Aepyornis maximus*.

Of especial interest, however, was that this gigantic bird, now known to have stood around 10 ft. tall and to have weighed almost half a ton, was already familiar to the local tribes. They called it the vorompatra or vouronpatra ("marsh-dwelling bird"), confirmed that this was also the bird that laid the gigantic eggs, and claimed that although very rarely seen nowadays, it was still alive!

Such claims substantiated the testimony of Admiral de Flacourt, who had briefly referred to the selfsame bird two centuries earlier in his Madagascan tome:

> Vouronpatra, a large bird which haunts the Ampatres [marshland area] and lays eggs like the ostrich's; so that the people of these places may not take it, it seeks the most lonely places.

Following the discovery of the elephant bird's eggs and bones, scientists postulated that although it had been flightless with only vestigial wings and resembled a sturdy ostrich rather than a flying eagle, sightings of this enormous species by travelers during the Middle Ages may have been sufficiently awe-inspiring to have given rise to the legend of the roc. Certainly Madagascar's great elephant bird (it is known by this name in order to differentiate it from various smaller, related species whose demised remains have since been disinterred here too) appears to have still existed at the time of de Flacourt, and may even have persisted in the more remote swamplands until the early 1800s.

With the subsequent deforestation of large swathes of Madagascar and the attendant climatic changes, yielding a major drying-out of marshes and swamps across the island, however, the great elephant bird's survival would have been gravely endangered. And if it was also extensively hunted by humans, there is little doubt that it would have been wiped out quite rapidly.

Nevertheless, during the late 1990s at least two independent searches were made for living great elephant birds, one by Czech explorer Ivan Mackerle, the other by Leicestershire conservationist Barry Ingram, but no evidence for this awesome species' survival was found. Tragically, scientists back in the early 1800s took too long to take notice of native testimony concerning the mysterious vorompatra, thus allowing the great elephant bird to vanish under their very noses—which is quite a feat for a 10-ft.-tall flightless bird to achieve!

Had they pursued such reports earlier, it is quite possible that they would have obtained some first-hand scientific documentation of living specimens. Instead, we have only its mighty skeletons and immense eggs in museums to study and gaze upon, silent witnesses to the erstwhile wonder of the astounding great elephant bird—one of the largest and most astonishing birds ever to have walked the planet, more amazing even than the legendary roc itself.

During recent years, a second major candidate for the roc's inspiration has come forward. In 1994, based upon subfossil finds on Madagascar, an enormous species of now-extinct bird of prey was formally described—*Stephanoaetus mahery*, the Malagasy crowned eagle. Believed to have died out by the early 16th century, this fearsome predator is thought to have fed upon giant lemurs (including specimens weighing as much as 26 lb) and possibly even upon the great elephant bird itself. If early European sailors visiting the island had observed such a spectacular eagle, it is easy to understand how exaggerated retellings of these sightings could have given rise to the roc legend.

OTHER MADAGASCAN MARVELS

In his book's coverage of Madagascan creatures of cryptozoology back in 1958, Bernard Heuvelmans wrote:

> In Madagascar fortunate circumstances have almost enabled us to watch the extinction of the giant fauna of the past, but we have missed our opportunity.

Although there seems little doubt that we are indeed too late ever to marvel at the sight of a living great elephant bird, judging from the reports presented here it is not entirely beyond hope that small numbers of one or more species of giant lemur and dwarf hippopotamus may still linger in the most inaccessible, least visited regions of this extraordinary island. Moreover, even the diverse assemblage of mystery beasts documented in this chapter is not the total cryptozoological complement that Madagascar can offer.

Others that have been reported down through the years include: a giant tortoise said to exist in various southwestern lake-containing caves that may constitute a surviving representative of the supposedly extinct species *Aldabrachelys* (=*Dipsochelys*) *grandidieri* (see Chapter 2 for more details); a giant bat-like entity termed the fangalabolo; an apparent horned mega-crocodile far bigger than any recognized by science and known as the railalomena (see Chapter 10); what may be a late-surviving representative of the officially long-demised giant aye-aye *Daubentonia robusta*; and a tiny, still-unidentified lemur known locally as the malagnira in western Madagascar's Tsingy de Bemaraha reserve.

Nor should we forget at least three specimens captured here of an enigmatic unclassified cat form resembling the mainland African wildcat *Felis silvestris lybica*; a mystifying but aggressive freshwater monster called the kavay; and even an extraordinary moth with a prodigious 15-in-long tongue whose existence has yet to be proven but which has been scientifically predicted on the basis of the presence in Madagascar of a particular orchid species, *Angraecum sesquipedale*, whose pollen can only be reached (and hence result in pollination for this plant) by a moth with a tongue of at least that length.

During the past two decades, many new species of lemur have been discovered here, including some large and very conspicuously colored examples, as well as two remarkable new species of large euplerid (Malagasy mongoose)—one boldly striped, the other semi-aquatic. So who can say categorically that there are not other, even more remarkable surprises still in store within Madagascar's long-isolated world of zoological endemics and enigmas?

Chapter 8:
YOU KNOW WHEN YOU'VE BEEN TRUNKOED —IT'S THE SURREAL THING!

> *In North America the black bear was seen by Hearne swimming for hours with widely open mouth, thus catching, like a whale, insects in the water. Even in so extreme a case as this, if the supply of insects were constant, and if better adapted competitors did not already exist in the country, I can see no difficulty in a race of bears being rendered, by natural selection, more and more aquatic in their structure and habits, with larger and larger mouths, till a creature was produced as monstrous as a whale.*
> — Charles Darwin, *On the Origin of Species* (1st edition only)

Back in early September 2010, my German colleague Markus Hemmler and I jointly solved one of the most perplexing of all cryptozoological mysteries (I initially documented the discovery in three world-exclusive ShukerNature blog posts). We revealed, almost 86 years after the remarkable events surrounding it had taken place, the much-speculated identity of Trunko and discovered no less than three hitherto-unpublicized photographs of its beached carcass. (I had light-heartedly dubbed the creature "Trunko" in my book *The Unexplained*, 1996, never suspecting for one moment that this would ultimately become universally adopted as its formal name!)

This truly bizarre entity has even been likened by some writers to Darwin's hypothetical evolved bear-whale (cited above). It was of course the infamous white-furred, proboscis-endowed sea monster from the early 1920s that had allegedly battled two whales out to sea at Margate, in Natal, South Africa, before its lifeless carcass had washed ashore on Margate's beach. There it had remained for 10 days before being carried back out by the tide, without ever having been examined by scientists or even photographed, and never to be seen by anyone ever again. That, at least, had long been the official Trunko story.

Representation of Trunko envisaged as a living trunk-bearing hairy cryptid (Lance Bradshaw)

However, in uncovering Trunko's identity and the photos, Markus and I also discovered that almost everything that had been written about this surreal specimen in the cryptozoological literature was wrong. Indeed, to put it bluntly, for the past 80-odd years cryptid investigators everywhere had been well and truly Trunkoed! So now, for the very first time in any cryptozoological book, here is the true, complete history of Trunko—or as true (or complete) as anything regarding such a creature of contradiction can ever be!

MAKING A DATE WITH TRUNKO

Even the year when Trunko had made its famous debut out at sea had formerly been unclear. The most commonly cited date for its whale battle and subsequent beaching had been November 1922 (or sometimes even as precise a date as November 1, 1922—as in *Living Wonders* by John Michell and Robert Rickard, for example), which is why this is what I too had previously given in my own publications documenting Trunko. However, several other dates had also been claimed.

According to veteran cryptozoologist Bernard Heuvelmans, for instance, the Trunko saga had occurred sometime *prior to* November 1, 1922. Conversely, a London *Daily Mail* newspaper report of December 27, 1924, gave the date in question as October 25, 1924 (misquoted as October 26, 1924, by Charles Fort in 1931 in his book *Lo!*). Mid-November 1924 was yet another date cited, this time in a *Wide World Magazine* article of August 1925, which consisted for the most part of a detailed account of Trunko penned by Johannesburg photographer A.K. Jones.

Moreover, it was our very discovery in September 2010 of that long-overlooked article containing Jones's account and two of his three equally-neglected published photographs that enabled me to identify Trunko and reconstruct the latter's history. For until then, no-one had realized that any Trunko photographs existed, or Jones's published first-hand eyewitness account and examination of this entity—which is why the cryptozoological world had only been able to speculate blindly concerning its identity and history.

TRUNKO—THE OFFICIAL BIOGRAPHY

Before presenting the principal text of Jones's rediscovered report, however, let us first examine three key Trunko accounts that have collectively yielded what until now had been widely considered to be the standard version of events concerning this maritime enigma.

First and foremost is the report referred to by Charles Fort that had appeared on December 27, 1924, in London's *Daily Mail* (referred to hereafter as the *Daily Mail* unless specified otherwise). As this classic account contains many of the basic elements of the Trunko saga's standard version, yet had never been quoted in full before within the cryptozoological literature, I am doing so now:

FISH LIKE A POLAR BEAR.
A FIGHT WITH TWO WHALES.
ESCAPE AFTER 10 DAYS' SLEEP.

"On the morning of October 25 I saw what I took to be two whales fighting with some sea monster about 1,300 yards from the shore. I got my glasses, and was surprised to see an animal which resembled a Polar bear, but in size was equal to an elephant. This object I observed to back out of the water fully 20 ft and strike repeatedly at the two whales, but with seemingly no effect." This is an extract from a letter sent to a Natal newspaper by Mr. H.C. Ballance, Margate Estate, South Coast, Natal.

BODY LIKE A BEAR

The letter continues:

"After an hour the whales made off and the incoming tide brought the monster within sight, and I saw that the body was covered with hair 8 in. long, exactly like a polar bear's, and snow white."

Next morning, Mr. Ballance found the carcass lying high on the beach. He measured it and found it was 47 feet from tip to tail. The tail was 10 feet long and 2 feet wide, and where the head should have been the creature had a sort of trunk 14 inches in diameter and about 5 feet long, the end being like the snout of a pig. The backbone was very prominent, and the whole body covered with snow-white hair.

"For 10 days," continues Mr. Ballance, "this mass lay inert. On the eleventh day there was not a sign of the creature.

"I met some natives who told me that while fishing they had seen the monster out at sea, going up the coast, and that is the last we have seen of it."

Coincidentally, the end of the *Daily Mail* report was followed by a brief mention of another titanic sea-battle spied off the Natal coast at around the same time as Trunko's and previously reported in the *Daily*

Mail on December 16, 1924. However, this confrontation featured a whale and a giant squid, whose tentacles were clearly observed when it was later washed ashore. (Nevertheless, that incident has been confused with the Trunko case in some published coverage of the latter.)

The only major discrepancy in this early report from the modern-day standard version of the Trunko saga is its intimation that Trunko had not died, but that after being beached for 10 days it had made its way back out to sea. However, this supposed activity was not personally observed by eyewitness Ballance, merely claimed by some locals who spoke to him about the creature. Also, the *Daily Mail* report's wording is sufficiently imprecise to lend an alternative interpretation, i.e. that Trunko was merely being carried up the coast passively, by the sea—were it not, of course, for this report's unambiguous subtitle, "Escape After 10 Days' Sleep."

Having said that, some much later newspaper accounts have even alleged not only that Trunko wasn't killed by the whales but, rather, that the whales were killed by Trunko! However, as will be seen, none of the key Trunko accounts includes such a dramatically conflicting claim, which can instead, I feel, be satisfactorily discounted as mere journalistic hyperbole.

Ballance's account was summarized and repeated in many subsequent media reports worldwide during the following 12 months or so, but not everybody accepted its veracity. Charles Fort was particularly skeptical. After devoting just two sentences to the *Daily Mail* report, he dismissed the entire subject as follows:

> I won't go into this, because I consider it a worthless yarn. In accordance with my methods, considering this a foolish and worthless yarn, I sent out letters to South African newspapers, calling upon readers, who could, to investigate this story. Nobody answered.

The second key account concerning Trunko was the coverage accorded it by Bernard Heuvelmans, who documented this seemingly unclassifiable marine creature as follows within his seminal tome *In the Wake of the Sea-Serpents* (1968). Curiously, however, as will be seen, Heuvelmans's account and the *Daily Mail* report disagreed with one another on a number of important issues, and even Heuvelmans's source for Ballance's testimony was a report in an unnamed South African newspaper, not the London *Daily Mail* report:

On 1 November 1922, three years after he bought the farm of Margate in Natal which has since become the seaside resort of that name, Hugh Ballance told the South African press a very strange tale:

"I saw what I took to be two whales fighting with some sea monster about 1,300 yards from the shore. I got my glasses and was amazed to see what I took to be a polar bear, but of truly mammoth proportions. This creature I observed to rear out of the water fully twenty feet and to strike repeatedly with what I took to be its tail at the two whales, but with seemingly no effect."

A gigantic Polar Bear with a tail long enough to use as a whip is clearly no ordinary bear—especially as the Polar Bear is never found in the southern hemisphere—it is also something quite new even for the most incorrigible of sea-serpent hunters.

The battle lasted three hours and was watched by crowds on the shore, after which the two whales made off, leaving the monster floating lifeless on the surface.

The next night the tide threw the great carcass on the beach. It was described thus by T.V. Bulpin in his book, *Your Undiscovered Country* [1965], about the beauties of South Africa:

"It was certainly a giant of a creature, forty-seven feet long, ten feet in breadth and five feet high. At one end it had a trunk-like appendage about fourteen inches in diameter and five feet long. The creature was covered in snow-white hair and seemed to be devoid of blood."

This incredible carcass lay on the beach for ten days. Many people came to stare at it. But no zoologist took the trouble to examine and identify it before the spring tide washed it away, for it has never been classified.

Heuvelmans's account introduced several elements that subsequently became fundamental components of the standard version of the Trunko saga. (Some of these, moreover, were mutually exclusive with, and directly replaced, their respective, earlier equivalents present in Ballance's testimony.)

Namely: the change of Trunko's date of appearance from Ballance's claim of October 25, 1924, to an unspecified date prior to November 1,

1922 (even though it is evident from the *Daily Mail* report of December 27, 1924, that Ballance's sighting had taken place in October of that same year, *not* two years earlier); additional dimensions of the beached carcass, and its apparent lack of blood; the claim that the carcass attracted no scientific interest or examination; and, most significant of all, the statement that it was lifeless and was finally washed back out to sea by the tide (rather than being still alive and making its own way out to sea, as ostensibly claimed by the locals who had spoken to Ballance).

But were any of these marked deviations from Ballance's statement correct? Not until our own discoveries took place in September 2010 could this query be answered.

The third key Trunko account was penned by South African writer and illustrator Penny Miller in her book *Myths and Legends of Southern Africa* (1979). Although her coverage of events included the version of Ballance's testimony quoted by Heuvelmans in his book, it gave the date of Trunko's appearance and beaching as November 1, 1922 (rather than *prior to* November 1, 1922, as claimed by Heuvelmans).

However, far from repeating Ballance's claim that this entity did not die but made its own way back out to sea, Miller's account reiterated Heuvelmans's statement that it washed up dead onto Margate's beach and specified that the actual site was beyond the aptly-named Tragedy Hill. In addition, Miller's became the first key account to emphasize the stench of decomposition emanating from this beached carcass:

> As recently as 1922 a dead monster was washed up on the golden beach of Margate. It was a nine-day wonder until the hot weather accelerated decomposition and the stench made its weird bulk unapproachable... For ten days it lay on the beach; even a span of 32 oxen failed to move it! The stench became more and more putrid until finally, the spring tide had pity on the inhabitants of Margate, and overnight the carcass vanished, leaving nothing behind but a cloud of mystery and speculation.

Miller's book also saw the debut of her now-iconic, frequently-reproduced/adapted line drawing of a huge, hairy, limbless, elephant-trunked creature lying dead on the shore with a couple of onlookers standing beside it.

That, then, was the trio of key accounts from which the modern-day standard version of the Trunko saga had evolved—resulting in all manner of published and online speculation (not to mention a rich diversity of imaginative illustrations) relative to the possible existence (or otherwise) somewhere in the vast oceans of a still-undiscovered, yet zoologically-inconceivable species of sea creature resembling an enormous hairy polar bear, yet sporting the trunk of an elephant, but with no recognizable head, and a conspicuous absence of limbs too.

Or, to put it another way, a beast so outlandish that nothing even remotely similar was known from either the present day or the fossil record. (Little wonder, perhaps, that some critics of cryptozoology had simply dismissed Trunko as a journalistic hoax.)

TELLING IT LIKE IT WAS—A.K. JONES'S TRUNKO ACCOUNT

But now, in order to compare and contrast with those key accounts, it is time to unveil A.K. Jones's extremely informative and enlightening first-hand account of the Trunko carcass as directly examined and photographed by him. Published in the August 1925 issue of *Wide World Magazine* but inexplicably overlooked afterwards for more than eight decades(!), it provides a wealth of morphological details never previously accessed by cryptozoologists, as well as a previously unreported claim that after the carcass had been washed out to sea following 10 days lying on the beach, it was subsequently washed ashore again.

Before I present his description of the carcass, however, I should point out that Jones's account began with a short summary of Trunko's offshore battle and its subsequent beaching, but that the dates he gave for these events were only approximate, yet very intriguing. He gave the date of Trunko's beaching as "about the middle of November last" (i.e. mid-November 1924), and the date of Trunko's whale battle as "about three weeks previously" (i.e. the last week of October 1924). The latter date thereby corresponds with Ballance's testimony in the *Daily Mail* report. However, Jones's account is the only one to claim that a time period of about three weeks had elapsed between the battle and the beaching. All others had claimed that the beaching took place just hours after the battle.

A second noteworthy discrepancy was Jones's claim that "...a terrific struggle had been witnessed out at sea by residents of Margate, between what they took to be a whale and some other animal which they could not clearly distinguish." Again, Jones's account is the only one to claim the involvement of just one whale, and also that Trunko was not clearly discernible by its shore-based observers (though this might explain a lot!).

So is Jones's account the correct one in either or both of these instances? Alternatively, can its unexpected discrepancies with all other key accounts be safely ascribed to nothing more mysterious than an imperfect recollection by Jones of events that he had not personally experienced, and which in any case had occurred and been reported by the media a full year before he had got around to writing about them himself? As yet, we simply do not know.

However, what we do know, and what is much more important anyway, is what he did experience personally—the beached Trunko carcass:

> About the middle of December [1924] it was discovered that the monster had been washed up a second time, and was now lying on the rocks about three miles farther up the coast. I happened to be spending a holiday at Margate at the time, and secured three photographs of the creature, taken from different angles. No one here is able to identify it, but perhaps some of your scientific readers may be able to throw some light on its identity.
>
> As the photographs clearly show, the monster is covered with slimy hair about four or five inches long, under which lies what I should take to be a very tough hide. The whole thing has the appearance of a huge sheepskin that has been thoroughly soaked. The body seems to be composed of extremely firm flesh; there is little "give" in it when poked with a stick. There are probably bones in the monster, but no actual bone can be felt, as the whole thing is so firm that even if there is bone it cannot be distinguished by touch. However, I should say that there is a bony framework, or else the hollow which runs the whole length of the back, clearly shown in the first photograph, would not still be so clearly defined as it is, considering the rough handling to which it must have been subjected by the sea.
>
> At one end there is a round lump about two feet in diameter, which might be taken for a head, but there are neither eyes, mouth, ears, nor anything else visible. There are, moreover, no limbs, flappers, tentacles, tail, or any other features which would help to identify it. The measurements are as follows: About fifteen feet long, six feet broad, and two feet thick. The carcass, at the moment of writing, has already begun to decompose, and there can be little doubt that it is composed of flesh of some sort. Probably, if it is allowed to rot and the

skeleton becomes visible, it will be possible to identify it by this means.

Worth noting here is that whereas Jones stated that the carcass lacked a tail, Ballance's description of it as quoted in the *Daily Mail* report referred specifically to a 10-ft.-long tail (another newspaper report even described the tail as lobster-like). As Ballance had viewed the carcass when it was originally beached, however, whereas Jones had only viewed it after it had been re-beached, the tail had probably dropped off during the carcass's intervening period at sea. Supporting this theory is Jones's failure to include any mention of Trunko's eponymous proboscis either, which therefore must also have been lost by then, explaining why the carcass was now much smaller than when viewed earlier by Ballance.

Jones ended his article by speculating whether an underwater earthquake that had been recorded near Margate just before Trunko's appearance may have perhaps dislodged this entity from the ocean-bed and thrown it up to the surface of the sea, in a manner similar to what had occurred some time previously when a sea monster had been washed ashore on South America's east coast following a nearby seaquake. (Interestingly, it is certainly plausible that the Margate seaquake was responsible for the surfacing, sighting, and subsequent whale battle off the Natal coast of the giant squid—normally a deepwater species— noted by the *Daily Mail* at the end of its Trunko report.)

In his account, Jones noted that he took three photos of the Trunko carcass. Two were published alongside his account, but it was the third photo, the one that didn't appear there, which initiated the chain of discoveries made by Markus and me during autumn 2010 that finally unmasked Trunko. I have already described these in detail in my ShukerNature blog posts and in an Alien Zoo report for *Fortean Times*, but they can be summarized and updated here as follows.

TRUNKO IDENTIFIED AT LAST

In early September 2010, Markus drew my attention to a page devoted to Trunko (margatebusiness.co.za/index.php?option=com_content&view=article&id=64%3A) on the website of the Margate Business Association (MBA).Uploaded on October 19, 2009, and entitled "The Legend of Trunco" [sic], much of its information was nothing new and not entirely accurate either (the date of Ballance's viewing of Trunko's whale battle, for example, was given as November 2, 1922). However, there were two very notable exceptions. One was a brief extract from a letter about Trunko by Johannesburg photographer A.C. [sic] Jones, who was also credited there as the author of a *Wide World Magazine* article on Trunko from July 1925 (this again was incorrect; it was August 1925). And the other, to our great astonishment, was a black-and-white photograph snapped by Jones of Trunko's beached carcass!

A.K. Jones's above-mentioned photo of Trunko's beached carcass (A.K. Jones)

Moreover, I was shocked to realize that I'd already seen an extremely similar image. The *Daily Mail* report of December 27, 1924, documenting Ballance's eyewitness testimony had contained a very small sketch of a long but largely featureless white carcass on a beach with the silhouette of a human figure standing in front of the carcass at its extreme left-hand end, looking down at it, with one arm outstretched. This sketch had simply been labeled "A sea monster," and so provided no hint that it may be directly related to the Trunko report.

However, looking at Jones's photo of the Trunko carcass, it was immediately obvious that the sketch had been based directly on this photo, because every discernible detail was the same—the carcass's shape and pale color, the skyline, the precise location, shape, and orientation of the human figure, even the angle of the figure's outstretched arm.

This was clearly no coincidence, and as Jones had snapped his photos during the same month (December 1924) as the *Daily Mail* report had been published (near the end of that month), I can only assume that the *Daily Mail* had somehow seen this particular Jones photo somewhere and had prepared a sketched version of it for inclusion in its report. Confirmation of this notion may well exist within a statement on the MBA's Trunko page that Jones's Trunko account had been published not only in the *Wide World Magazine* but also in the *Rand Daily Mail* (a Johannesburg newspaper).

No date of publication was mentioned for this *Rand Daily Mail* report (which no doubt included Jones's photos as well as his account), and neither Markus nor I have so far been able to trace it. However, if this report had been published prior to the publication of the London *Daily Mail*'s own, separate Trunko report (which is more than likely, given the much greater newsworthiness of a South African sea monster to a South African newspaper than to a London one), then it well may be that the *Rand Daily Mail* report would have been seen by at least a few reporters and artists at London's *Daily Mail*. In turn, this may conceivably have inspired the latter newspaper to produce its own Trunko report, complete with a sketch of a photo from the *Rand Daily Mail*'s version.

Although, prior to my recognition of the MBA's Jones photograph as the model for the London *Daily Mail* report's sketch, we both naturally wondered whether this potentially explosive photo could be a hoax. I could immediately see that the carcass as depicted in it bore a striking resemblance to various classic hairy globsters that have been reported from beaches all over the world down through the decades.

The zoological identity of these huge, amorphous, hairy masses formerly incited considerable controversy in scientific circles, but recent DNA analyses of tissue samples taken from various specimens have confirmed that a globster is merely a massive, tough skin-sac of blubber containing collagen (and occasionally an isolated bone or two) that is sometimes left behind when a whale dies and its skull and skeleton have separated from the skin and sunk to the sea bottom. Moreover, this skin-sac's external surface is usually covered in exposed connective tissue fibers that resemble pale, shaggy, scraggy hair or fur, and there is no trace of blood as this has long since drained or been washed away.

Despite numerous attempts, Markus and I had both failed to elicit any reply from the MBA concerning that remarkable photograph on its website's Trunko page (or even whether Jones was still alive and contactable, bearing in mind that even if he had only been a late teenager back in 1924, he would now be a centenarian).

Consequently, the only option remaining for determining whether it was a genuine image of the Trunko carcass was to locate Jones's *Wide World Magazine* article, and see if it contained this or any other photos (or at least a detailed description) of the carcass. Happily, within the space of just four days Markus and I had independently succeeded in doing precisely that. In my case, paranormal writer and researcher Richard Holland kindly provided me with a copy from his extensive collection of *Wide World Magazine* back issues. And the rest, as they always say, is history.

There before us were the two additional Jones photos and his in-depth description, confirming beyond any doubt that the Trunko carcass had indeed been a globster.

As for its famous trunk, this was probably an isolated bone encased in blubber (or else a long tubular evagination of blubber, or possibly even the remains of a throat pleat if the pre-globsterised whale had been a baleen species). But as this meant that Trunko had never been alive in the sense of being an exotic, trunk-bearing, snowy-furred mystery beast, how can its lively battle out to sea with the whales be explained?

In fact, long before I had discovered Jones's photos and article, I had already documented what I believed to be the answer to this riddle. While preparing the Trunko section for my book *Extraordinary Animals Revisited* (2007), I had come across a meticulous examination of the Trunko phenomenon that had been undertaken and posted by American cryptozoological investigator Lance Bradshaw on his Kryptid's Keep website (no longer online).

In his account, Lance had postulated with great ingenuity and originality that the Trunko battle could be reasonably explained as an optical illusion. That is, observers on the shore looking some distance out to sea thought that they were watching some bizarre furry mega-beast battling two whales, but what they were really seeing was two whales repeatedly throwing into the air a huge but already dead carcass,

A.K. Jones's two additional Trunko photos (A.K. Jones)

playing with it in an animated manner already on record for certain cetaceans (particularly killer whales, which are indeed native to South Africa's coast). It is even possible (as I've already proposed for the *Daily Mail*-reported giant squid) that this carcass had originally been propelled from the sea depths up to the surface by the recent Margate seaquake mentioned by Jones.

And now, with our shattering discovery that Trunko as a living snowy-furred, trunk-sporting cryptid had never existed, and that in reality it had been nothing more than a long-dead, globsterised whale carcass, Lance's theory had finally been confirmed. Needless to say, this also explained why the body of a supposedly battle-scarred, fatally wounded cryptid had not been pouring (or at least copiously stained) with blood.

Finally, therefore, the true history of Trunko would appear to be that in late October 1924, various onlookers present on Margate's beach had observed at least one whale (but most probably two) some distance out to sea boisterously playing with a globsterised whale carcass that was later washed ashore onto the beach. There it lay for 10 days, decomposing very odiferously before being carried out to sea by the tide, then washed ashore again in December 1924, where it was closely observed, photographed, and examined by A.K. Jones before the tide took it back out, after which it was not reported again.

But Trunko would not of course be Trunko without leaving behind some still-unanswered questions. In particular, how can the notable discrepancies between the details in Ballance's account and those in Heuvelmans's be explained? Having said that, however, now that Trunko has been conclusively exposed as a globster, it is evident that Ballance's claim that it made its own way out to sea again after having been beached earlier was nonsense. For almost a century, it may indeed have acquired the mystique of a veritable maritime mirabilis, but even the indefatigable Trunko was incapable of resurrecting itself from the dead!

ALASKA'S "SON OF TRUNKO"?

While investigating Trunko, Markus also solved a longstanding mystery concerning the identity and whereabouts of a smaller but no less intriguing white-furred, long-snouted sea monster carcass. Discovered washed ashore on Alaska's desolate Glacier Island, the discovery date of this "Son of Trunko" was traditionally thought to have been November 1936, but was later found to have been November 10, 1930.

After it had been examined by a scientific team, cryptozoologists had traditionally assumed that the creature's remains had not been not preserved and that its zoological identity remained a mystery. Markus, however, discovered not only that it had indeed been formally identified as a minke whale *Balaenoptera acutorostrata*, but also that its complete skeleton had eventually been donated to the National Museum of Natural History in Washington DC, where it remains to this day, officially labeled as USNM 256498.

TRUNKO STOP PRESS—A FOURTH PHOTOGRAPH IS DISCOVERED!

In March 2011, fellow Trunko investigator Bianca Baldi informed me that she had uncovered a hitherto-unpublicized Trunko photograph. This latest, fourth photo was discovered by Bianca (whose home town actually happens to be Margate, South Africa) in the archives of Margate Museum, contained within a document entitled "Margate Area Annals 1917 to 1969 compiled by the president and members of the Margate Women's Institute." As with the previous three, it clearly reveals Trunko to have been a globster and not some exotic undiscovered species of white-furred, trunk-brandishing sea beast as once believed—or at least hoped for—by many cryptozoological investigators.

ADDENDUM—THE TRUNKO PROBABILITY

In *Fortean Times* #277 (July 2011), reader Tony Venables responded to my revelatory feature-length article on Trunko in the May 2011 issue and challenged my assertion that the rebeached carcass seen and photographed by A.K. Jones was one and the same as the original beached Trunko carcass. I replied via a detailed letter defending my assertion, which was published in slightly edited form in *FT* #279 (September 2011); but here, in print for the very first time anywhere, is the original, unedited version of my letter:

The fourth Trunko carcass photograph (photographer unknown)

> In his published letter (*FT*277), Toby Venables states that because A.K. Jones did not observe or photograph the Trunko carcass when it was originally beached ashore at Margate, South Africa, in mid-November

1924 but only when it was rebeached there about a month later in mid-December, my solving of the longstanding Trunko mystery—identifying the Trunko carcass from Jones's recently-rediscovered photos as having been a beached globster—can only be said to be probably correct, not definitely so, as the rebeached carcass may actually have been a different one from the original Trunko carcass (especially as it was smaller than the original one). Naturally, this possibility cannot be entirely discounted. However, as I'd now like to demonstrate, it is one so remote as to be unworthy of serious consideration.

Let us play devil's advocate for a moment, and thus adopt Mr Venables's hypothesis, that the original (November) and rebeached (December) carcasses were indeed separate entities. Yet apart from the smaller size of the December carcass, and its lack of the trunk and tail that were reported for the November one, the two are extremely similar. That is, they were both limbless, headless, shapeless, bloodless, covered in a very striking 'pelage' of snow-white hair, and highly decomposed. Bearing in mind that if any beached, decomposing carcass is subsequently washed back out to sea, chunks of it will definitely fall away and/or be ripped from it by marine scavengers or by predators such as whales or sharks, especially elongated well-delineated portions such as the trunk and tail (which would have had only a relatively small, fragile attachment site to the main mass), these differences between the two carcasses are readily explained.

Far less plausible, conversely, is the prospect that two completely separate carcasses sharing such unusual, distinctive characteristics as those noted above had washed ashore on the same coastline within a month of each other, particularly when (at least to my knowledge) no similar carcasses had ever done so there before, or indeed since.

And even if, by some exceptionally unlikely fluke, two such carcasses had genuinely done so, it doesn't actually make a great deal of difference to the likelihood of my identification of Trunko as a globster being correct. For many decades, globsters were a major cryptozoological mystery. In the past few years, however, the identification of what a globster truly is has finally been ascertained, using DNA analyses of tissue samples obtained from various recently-beached globsters. In every case, the globster has been shown to be an amorphous, limbless, headless, bloodless mass that constitutes a tough skin-sac of decomposed blubber containing collagen, whose exposed fibres yield the snowy 'fur' so frequently reported. Occasionally, an isolated bone or two with associated tissue have also remained within this mass, superficially resembling tentacles or other elongate organs. The above description corresponds perfectly with descriptions of the original Trunko carcass, including its initially anomalous head-lacking trunk and tail.

As for the Trunko scenario of a beached globster subsequently being washed back out to sea and then being rebeached, this is hardly unique. As recently as June 2011, for example, precisely the same series of events occurred with a globster washed ashore on the West Indian island of Barbados. And way back in autumn 1808, it had also happened with the famous Stronsay 'sea serpent'—in reality the carcass of a very large basking shark. Of course, Mr Venables may well reiterate that we cannot be sure in either of these cases that it was the same carcass that was rebeached. However, to adopt this attitude is to multiply improbability by improbability, whereas I prefer to adhere to Occam's Razor—namely, the simplest solution to a problem is more likely to be the correct one, provided that it satisfactorily covers all of the facts. And I believe that my Trunko solution—featuring a single globster carcass that was originally beached and subsequently rebeached—does.

After all, the alternative is that lurking somewhere out there in the ocean depths off South Africa, unseen by the local inhabitants and still-undiscovered by science, is a population of totally outlandish, white-furred, limbless creatures brandishing a lengthy tail for whipping whales with, as well as an elephantine trunk, yet lacking any recognisable head, and thus constituting a species wholly unlike any other animal form either alive today or known from the palaeontological record. And just how probable is that?

Chapter 9:
TOADS IN HOLES AND FROGS IN THROATS — THE ENIGMA OF ENTOMBED LIFE

Like a toad within a stone
Seated while time crumbles on;
Which sits there since earth was curs'd
For Man's transgression at the first;
Which, living through all centuries,
Not once has seen the sun arise;
Whose life, to its cold circle charmed,
The earth's whole summers have not warmed;
Which always—whitherso the stone
Be flung—sits there, deaf, blind, alone;

— Dante Gabriel Rossetti, *Jenny*

The phenomenon of toads, frogs, and sometimes other creatures (lizards, insects, worms, and at least one bat) discovered entombed yet allegedly alive inside solid rock is one of the strangest curiosities of unnatural history. More than 200 cases have been documented, with many more undoubtedly known, spanning many centuries though attaining a peak in popularity during the 1700s and 1800s. Yet despite attempts by science to provide mainstream solutions, it remains a somewhat atavistic anomaly, a throwback to those Victorian times when two-headed sideshow freaks, stuffed mermaids, and other zoological oddities were à la mode, and nowadays sitting somewhat uneasily and still only incompletely resolved in the modern high-tech world. Moreover, comparably bizarre scenarios, featuring organisms sealed inside coal, wood, plaster, ice, and even weirder media have also been documented, as revealed here.

WHEN THE STONE WAS OPENED...

Contrary to most accounts of immured amphibians and suchlike, which claim that the earliest cases on record date from the 16th century, the phenomenon had been recorded well before then. Indeed, the first example I know of can be found documented in a chronicle from 1145 by Robert of Thorigny, in which he referred to a living toad that had been discovered enclosed in a stone within the ramparts at Le Mans, France. A similar case was recorded 53 years later by William of Newburgh in Book 1 of his *Historia Anglia* (1198).

By the 1500s, however, the subject had indeed become one that was frequently documented, and by such eminent chroniclers and scholars as Aldrovandus, Agricola, Simon Majol, Martin Delrio, and, most famously, France's celebrated royal surgeon Ambroise Paré:

> Being at my seat near the village of Meudon, and overlooking a quarryman whom I had sent to break some very large and hard stones, in the middle of one we found a huge toad, full of life and without any visible aperture by which it could get there. The labourer told me it was not the first time he had met with a toad and the like creatures within huge blocks of stone.

Paré believed that such creatures must have generated spontaneously from mud or slime within their stony cells. And certainly, when reading through the most detailed accounts of entombed animals, one can well understand his bewilderment as to how they had come to be incarcerated in such an all-encompassing manner. Take, for instance, the following selection.

In September 1770, at Le Raincy Castle, France, a live toad was found encased in plaster during the demolition of a wall that had stood for at least 40 years. This case was personally investigated and confirmed by Jean Guéttard, a Fellow of France's Academy of Sciences.

In 1835, John Bruton saw a slab of sandstone drop off a wagon that was running down to the embankment at a railway cutting near to Coventry in the West Midlands, England. When the slab hit the ground, it broke across the middle, revealing a hole at its center—and falling out of that newly exposed hole was a live toad. Tragically, the toad hit the ground with such force that it damaged its head, but after Bruton picked it up it nevertheless survived for a further 10 days on display at his office workplace before ultimately expiring.

On September 20, 1852, a living toad was found inside a spacious, crystal-lined cavity when a large, heavy lump of iron ore was split open by quarrymen at Paswick, Derbyshire, after digging it up from a depth of around 13 ft. beneath the surface of the ground. Once released,

19th-century engraving depicting the discovery of a toad inside a rock that had been split apart (public domain)

however, the toad died almost immediately.

In 1862, London's Great International Exhibition displayed a big block of coal obtained from a seam deep beneath Newport, Monmouthshire, and next to this was a living frog that the seam's miners claimed to have found inside it. Several other incidents of frogs and toads found inside coal have also been documented. These include, most memorably, one that featured the expulsion of a mummified *Hyla* tree frog from inside a large exploding lump of coal, as witnessed by teenager Eddie Marsh of McLean County, Pennsylvania, during the early 1880s. This mystifying specimen was later given to James Stevenson of the U.S. Geological Survey.

One of the most dramatic cases on record occurred in the vicinity of Uitenhage, South Africa, during autumn 1876. Timber men sawing a 16.5-ft.-wide tree trunk into planks had just removed the bark and the first plank when they spotted a hole the size of a wine glass—inside which were almost 70 tiny grape-sized toads. According to the men, they were:

> ...of a light brown, almost yellow colour, and perfectly healthy, hopping about and away as if nothing had happened. All about them was solid yellow wood, with nothing to indicate how they could have got there, how long they had been there, or how they could have lived without food, drink or air.

More recently, a team of railway workers extending a track near Te Kuiti, New Zealand, in 1982 was trimming through a layer of mudstone, a type of sedimentary rock, when one of the men revealed a small cavity in the rock, 13 ft. beneath the surface of the ground. As he peered closer, he was amazed to observe a living frog ensconced inside it. Later, another cavity within this same layer of rock was also found to contain a living frog. One of the eyewitnesses to these enigmatic finds was the railway's works supervisor, a Mr. L. Andrews, who confirmed that it was impossible for the frogs to have fallen into the holes during work on the track (an explanation often put forward by skeptics of the entombed toad phenomenon) —they were definitely already inside them when the rock was split open during trimming.

A particularly noteworthy case was posted on Christmas Eve 2003 to an online message board devoted to mysteries by a lady with the screen name "TheQuixote." Having assisted in felling a Scots pine, she had been chopping at its trunk with a billhook when she saw something move where she had just cut the trunk. After scraping the area with the hook and ripping off some bark, she was startled to discover a live adult grey toad inside a cavity hitherto completely sealed by the overlying bark. She duly extricated the toad, which was somewhat damp but seemed healthy in spite of its enforced confinement. Prior to the tree's felling, the cavity in which it had been entombed would have been at least 6.5 ft. above the ground.

On November 1, 2011, Peter Horler of Bridgnorth in Shropshire, England, emailed me the following details concerning his recent discovery of an entombed toad. This interesting case has not previously been published.

> A few years ago now I was working in a quarry with a digger and lifted an 8 inch diameter sphere out of rock and put it to one side. Later that day at lunch break I cracked it open to find a small toad inside. Its eyes and mouth looked completely sealed and I thought it was dead. Then after about 25 minutes as I sat eating my sandwiches in the sun I caught something moving in the corner of my eye. The toad was moving about for about 5 minutes and then it stopped...I know little about Geology but these calcite type spheres come out of the rock (sandstone, limestone mud rock that lifts up as it shears horizontally when dug). It is impossible for me to remember the location of the quarry other than to say that it was in the West Midlands area because I have worked in so many locations over the years. I think it could have been in 2004...it was approx 6 metres [20 ft] under ground level.

IMMURED IMPOSTERS

Although there have been many cases reported, there is currently only a single-known preserved example of an entombed toad—and, regrettably, this also happens to be the most controversial case on file. On display at the Booth Museum's Cabinet of Curiosities in Brighton, England, it consists of a hollow flint nodule, inside which is the mummified corpse of a toad. It was allegedly found inside when the nodule was cracked open by workmen in a quarry at Lewes, East Sussex, in 1899. Not aiding its chances of being authentic, however, is the disquieting fact that in 1901 this intriguing specimen was donated to the Booth Museum by none other than Charles Dawson—not only the prime suspect behind the notorious Piltdown Man hoax but also someone who is known to have carried out a number of other scientific frauds and plagiarisms.

On April 7, 1865, workmen excavating in Hartlepool, England, split open a large block of magnesium limestone, and inside they discovered a living toad within a cavity just large enough to contain its body and such a snug fit that press reports stated that the cavity looked as if it had been cast around it. The toad's eyes allegedly shone with some brilliancy but its mouth was sealed, despite which it was able to produce a series of barking sounds whenever it was touched, reputedly via its nostrils. The toad became the property of the Rev. Robert Taylor of Hartlepool's St. Hilda's Church, and it attracted considerable interest—until a team from the Manchester Geological Society examined the cavity in which the toad had been sealed inside the block and discerned the marks of a chisel! Following suggestions that one of the workmen had perpetrated a hoax (and, when challenged, he did not deny this claim), the Hartlepool entombed toad's scientific appeal diminished as rapidly as it had arisen.

Moreover, during the 1800s a group of decidedly crafty craftsmen became notorious for baking toads into clay pots in order to sell them for a tidy profit to gullible naturalists!

Mention, albeit of the dishonorable kind yet again, must also be made of the infamous yet tenacious yarn telling of a living pterodactyl that in 1856 supposedly staggered out of a huge block of Jurassic limestone blasted apart by a team of workers excavating a railway tunnel at Culmout, France. Its species was identified as *Pterodactylus anas* a fictitious zoological name, with *anas* translating as duck...or, in French, canard!

A CANNY EXAMPLE?

Giving the subject of incarcerated amphibians, a decidedly modern-day slant is the remarkable case of the frog and the coke can. In 1995, while clearing away some rubbish close to a pond in the quadrangle of Rhyl High School in Wales, schoolboy Paul Astbury was very surprised to find a frog squatting inside a discarded Cherry Coca-Cola can—because, as confirmed by David Wilson, the school's head of science, the frog was so much bigger than the can's ring-pull hole that if its discoverers had not released it by cutting the can open, it would never have been able to get out. So how had it originally got in?

DIGGING UP A SOLUTION

As with so many other longstanding mysteries, several very different solutions to the enigma of entombed animals have been suggested down the centuries. During the pre-evolution age, Paré's notion of spontaneous generation was widely supported, but by the early 1800s the subject had begun to attract formal scientific attention. This was epitomized by the meticulous experiments conducted in 1825 by the Rev. William Buckland, Professor of Geology at Oxford University, who sealed a series of toads of differing sizes inside identical circular cells, some within blocks of porous limestone, others within blocks of non-porous sandstone, and all buried under three feet of earth within his garden.

A year later, he dug up the stone blocks and examined the toads sealed inside them. All of those interred within the sandstone blocks were dead, as were the small toads within the limestone blocks. But the larger toads within the latter blocks were all still alive and two had even gained weight. After reburying the survivors inside their limestone cells, he examined them periodically during a second year, but by the end of this one they too had died. So, did these experiments thereby disprove allegations of entombed animals surviving for considerable periods of time? Not necessarily.

Today, the most widely-accepted explanation for such survival is that the toad or frog enters the rock (or coke can!) when still only very young and small, through a pore (as in rocks such as limestone) or via a channel (or ring-pull hole) leading inside, and then becomes trapped if it grows any larger before attempting to get out again. Its own smell may be enough to attract tiny insects inside, on which the amphibian will feed; air can easily penetrate through the rock's pores or channel and be absorbed directly through the amphibian's porous skin as well as inhaled into its lungs; and rain seeping in will also sustain it during its enforced encapsulation. And if it enters a state of torpor, so that its metabolic rate plummets, as happens when toads hibernate during adverse environmental conditions, this will enable it to survive for a much longer period than if it remains in its normal, fully active state.

Consequently, as critics of Buckland's experiments noted, regularly re-examining the toads as he did would have disturbed them and thus prevented them from entering hibernation, during which state they might have survived for much longer periods than when permanently awake.

Nevertheless, even this "official" solution fails to explain those accounts of creatures found alive embedded in solid blocks of coal, ore, etc, that lack any pore or channel—always assuming, of course, that these accounts are genuine.

Having said that, certain reports may indeed have a comfortingly conventional zoological explanation. During the 1990s, *New Scientist* published some hitherto-undocumented accounts, including the finding of a live frog inside a rock broken open by a sapphire miner while working just outside Adelaide, South Australia. The frog died almost immediately, so he took it to the state museum, along with a few other rocks of the same type, whereupon the museum staff duly found a frog in one of those, too. However, these could well have been water-holding frogs *Litoria platycephala*. Often inhabiting arid regions, during prolonged dry periods frogs of this native Australian species can bury themselves in mud and aestivate (summer hibernation) for months at a time, each sealed inside a specially secreted membrane to reduce water loss that hardens until a rock-like capsule is created. The sapphire miner may have dug up some of these capsules and mistaken them for genuine rocks.

The Australian water-holding frog (public domain)

Conversely, a number of decidedly esoteric theories have also been proposed for animal entombment—involving such paranormal-inspired concepts as the teleportation (possibly through time as well as space) of toads inside cavities within solid rock (but why, and by whom?); the existence according to Buddhist teachings of "occasional hells" overlapping our own world in which creatures are enclosed awaiting redemption; and tiny pocket-realms inside rocks that have been "overlooked" or shielded by time where an incarcerated toad could live indefinitely without ageing or requiring nutrients. This last-mentioned notion has been used to explain certain dubious reports of prehistoric newts, lizards, and other small non-pterodactylian(!) creatures having emerged alive from newly revealed cavities in rocks dating back thousands if not millions of years.

THE ICE NEWTS COMETH!

Living creatures unaccountably entombed in ice rather than rock have also been documented from time to time, but they were generally disregarded as fraudulent or mistaken—until 1988. This was when scientists at far-eastern Russia's Magadan Institute of Northern Biology announced that they had successfully revived an Asian salamander of the genus *Hynobius* that had been frozen for 90 years at a depth of 33 ft. within permafrost ice in Siberia's Kolyma region by placing it into a tub of cold water.

Normally this amphibian only has a lifespan of about 10 years, but inside its disproportionately large liver it stores glycogen, which, when external temperatures plummet, transforms into glycerine and spreads throughout the animal's body, acting as antifreeze and thus preventing its tissues' cells from crystallizing. It can then remain in this suspended state for remarkably long periods of time until the external temperature rises again.

Prior to this, there had been various Soviet claims of newts restored to life after having been retrieved from permafrost layers dating back many millennia, but these were never confirmed. True, one living newt was indeed obtained from such a layer, but it had apparently found its way there by falling through a recent crack in the permafrost; it hadn't been trapped there when the latter had originally formed.

SEEDS OF THE ANCIENTS?

Another oft-told tale of entombment features plant seeds that have germinated after having remained in dormant seclusion for many centuries if not millennia. Such claims have featured barley seeds from the 3000-year-old tomb of the Egyptian boy-king Tutankhamen, submerged lotus seeds found in a 3000-year-old boat near Tokyo, and arctic lupin seeds associated with rodent burrows shown to be 14,000 years old. Generally, however, when the data has been rigorously examined scientifically, the seeds' alleged antiquity either has been disproved or has raised serious doubts—with one notable exception. In 2002, an international team of botanists documented a lotus fruit germinated from a carbon 14-dated 1300-year-old seed-containing fruit recovered from an originally cultivated but now dry lake bed in northeastern China. This duly became the oldest formally verified germinated fruit known to science.

FROM PEARL-EMBEDDED FISHES TO BASILISK EGGS

They may be the most frequently reported, but stone- or rock-entombed toads are by no means the only cases on file of unexpected animal immolation—as the following intriguing selection demonstrates.

In 1886, within the *Proceedings of the Zoological Society of London*, German-born British zoologist Albert Günther documented an extraordinary item from his own personal collection of zoological specimens. It was an old shell belonging to the common pearl mussel *Margaritifera margaritifera*, a species of bivalve mollusk, embedded inside which was a small fish that was completely sealed inside a thin layer of nacre (mother of pearl). Indeed, the nacre had preserved this fish so precisely that not only its body's general outline but even its eyes and mouth could be readily discerned.

The fish was of a species often found living inside bivalves like that one, but this particular specimen had evidently penetrated too far inside the mollusk's shells, physically irritating the mollusk in just the way that a piece of grit or some other foreign body would do. Consequently, the mollusk had responded in precisely the same way that it would in relation to a foreign body—it had rapidly enveloped the fish inside a secretion of nacre, thereby converting it into a fish-shaped pearl!

As noted earlier in this chapter, it is possible that its porous skin would assist in keeping alive an entombed amphibian such as a frog or toad by enabling it to obtain oxygen via direct absorption through its skin as well as by normal pulmonary respiration. The same cannot be said of a scale-covered reptile, however, which is why the following case of a supposed entombed snake is so interesting. While browsing through the March 8, 1823, issue of a bygone English journal entitled *The Mirror of Literature, Amusement, and Instruction*, I noticed the following account of a very remarkable reptile:

> The newspapers of June 1772, state, that a living adder [*Vipera berus*] was found in a block of stone of 30 French feet diameter, the centre of which it occupied. It was twisted nine times around itself in a spiral line; it could not support the weight of the atmosphere, but died in a few minutes after it was taken from the stone. On examining the stone, not the least crevice could be discovered through which it might have crept, nor the minutest opening through which it could have received fresh air, or inhale any sort of sustenance.

To explain the adder's incarcerated existence via traditional belief, we must postulate that the stone initially possessed an opening sufficiently large to permit the entry of a 5-to-8-inch-long snake (the size of a young adder) before eventually becoming sealed in some way and thereby entombing the hapless animal indefinitely. And if the snake were already an adult, an opening large enough to facilitate the passage of a 20-to-24-inch-long snake would have been required.

Even more perplexing is how any snake, young or adult, could respire while so interred, because scaly reptiles like this cannot respire through their skin. And how could it feed? Perhaps it didn't; perhaps it underwent a prolonged period of torpor, comparable to the enforced rest embarked upon by snakes during the cold winter months in temperate climates.

Only one thing, in fact, *is* certain. Assuming the report to be genuine, more than 200 years after the finding of this serpent in stone, science is still unable to provide a conclusive explanation for it. In contrast, an answer may be forthcoming from folklore.

As described in Chapter 6, in pre-scientific times the fossilized spiraled shells of those prehistoric squid-related mollusks known as ammonites were popularly deemed to be the petrified remains of snakes, from which their heads had dropped off. Accordingly, it is quite possible that this strange case of an alleged adder imprisoned in stone was based upon nothing more substantial than a distorted account of a fossil ammonite discovered inside a stone block (with the claim that it lived briefly after its release from the stone block merely being journalistic license).

Well worth noting in support of this proposal is the close morphological correspondence between the flat, tightly-coiled spiral of many ammonites' shells and the report's description of the entombed adder as being "...twisted nine times around itself in a spiral line." Also worth remembering is that at the time of this report, the true zoological identity of ammonites had not been determined by science, thus increasing the likelihood for confusion with snakes.

And speaking of confusion with snakes: according to traditional European mythology, the basilisk was a small but deadly serpent that brought instant death to anyone unlucky enough to meet its gaze, and which hatched from an egg that had been laid by a cockerel and incubated afterwards by a toad. There are many reports from bygone days that tell of chicken eggs from which small serpentine creatures have emerged that were duly judged to be basilisks. A number of contemporary cases are also on file, but as the basilisk was wholly legendary, what could these modern-day egg-emerging snakes have been? The answer is as unexpected as it is unpalatable.

Chickens are often infected with parasitic gut-inhabiting worms, including the ascarid roundworm *Ascaris lineata*, a nematode species that can grow to a few inches in length. (A related giant species in humans can grow to more than 1 foot in length!) They are often passed out of the bird's gut when it defecates. Unlike in mammals, however, the bird's gut and its reproductive system share a common external passageway and opening—the cloaca. Sometimes, therefore, an ascarid worm ejected from the gut finds its way into the bird's reproductive system, rather than being excreted into the outside world, and moves into the oviduct. Once here, however, it becomes incorporated into the albumen of an egg, inside which it remains alive yet trapped when the egg is laid. But as soon as the egg is broken open to eat by some unsuspecting diner, the worm wriggles its way out of it to freedom, scaring the diner and perpetuating the myth of the basilisk in the process!

On March 3, 2012, Copenhagen University zoologist Lars Thomas left a message below a basilisk post on my Eclectarium of Doctor Shuker blog, informing me of his own direct experience with this fascinating phenomenon:

> I suppose you know, that some legends say young basilisk "worms" could sometimes be found in chicken eggs. Indeed if you found a worm in an egg, not so very long ago, old folks would tell you it was a basilisk, and tell you how to get rid of it. When I was 8 years old, I was on holiday at my aunt who had a small farm. One day I was helping her in the kitchen cracking eggs, and in one of them was a worm. Auntie told me it was a young basilisk, and that I should very carefully take it out in her garden and bury it, and then walk three times around the filled up hole. So I did, but not before making a drawing of the egg and the worm. I still got the drawing.

Judging from this telling little vignette, burying traditional belief in basilisks is clearly much harder to do than burying the supposed basilisk itself!

A FROG—OR MOUSE—IN THE THROAT?

If basilisk eggs are not grotesque enough, how about the following two examples of anomalous animal incarceration?

As noted in the *Veterinary Record*, in October 1888 a veterinary surgeon by the name of Mr. Wilkins was called out to the farm of a Mr. Willis of Newbury in the United States to shoot his valuable cart-colt, which had been suffering very severely for some time from a mysterious ailment that was greatly restricting its breathing. After the sad deed had been done, the colt's carcass was cut up, but when its neck was severed everyone was astonished to see a large toad crawl out of the windpipe, instantly explaining why the poor horse had been unable to breathe properly!

More recently, and again recorded in the *Veterinary Record*: after a race horse on a South African racecourse began coughing up blood following its winning of a race there, it was duly examined using an endoscope—which revealed the presence of a live mouse trapped inside its pharynx. A truly astonishing occurrence that surely gives the expression "winning by a narrow squeak" a whole new dimension of meaning!

A STUDY IN SCARLET?

And speaking of truly astonishing occurrences: whereas examples of toads and frogs discovered entombed alive in seemingly solid blocks of stone are fairly numerous, examples of other creatures incarcerated in solid materials are far less so. Consequently, I was very interested to receive from fellow anomalies investigator Richard Muirhead in November 2011, the following old newspaper report from the *Manchester City News*, which had been published in either March or April 1888:

> The torpidity in which bats remain through the winter season in the temperate and colder climates is well known, and in common with other animals undergoing the same suspension of powers they have their histories of long imprisonments. The following curious instances may serve to corroborate each other. A woodman engaged in splitting timber for rail posts in the woods close by the lake of Haming, belonging to Mr Pringle in Selkirkshire, discovered in the centre of a large wild cherry tree a living bat of a bright scarlet colour, which, as soon as it was relieved from its entombment, took to its wings and escaped. The recess in the tree was only just large enough to contain the animal; but all around the wood was perfectly sound, solid and free from any opening through which the atmospheric air could reach the bat. A man while occupied in splitting timber near Kelsall, Cheshire, discovered many years ago in the centre of a large pear tree a living

bat of a bright scarlet colour, which he foolishly suffered to escape from fear, being fully persuaded that it was not a being of this world. This tree also presented a small cavity in the centre where the bat was enclosed, but was perfectly sound and solid on each side. The scarlet colour of these prisoners is still inexplicable.

Indeed it is. However intriguing the mechanism responsible for their woody encapsulation may be, the anomalous coloration of these bats is even more so, at least to my mind. I hardly need point out that there is no species of scarlet-furred bat native to Britain, so how can their bizarre pigmentation be explained? Could it be that dye from its berries had somehow penetrated the cherry tree's wood and stained the bat encased within it? But what of the pear tree's bat. All thoughts or additional information would be greatly appreciated!

The last word on the thoroughly perplexing subject of live animal encapsulation belongs to an anonymous author writing just over a century ago in the scientific journal *Nature* for 1910, but which remains as true today as it was then:

One thing is certainly remarkable, that although numbers of field geologists and collectors of specimens of rocks, fossils, and minerals are hammering away all over the world, not one of these investigators has ever come upon a specimen of a live frog or toad imbedded in stone or in coal.

And that, perhaps, is the greatest enigma of all.

Chapter 10:
CROCODILIAN MONSTERS AND MYSTERIES

I have interviewed eye witnesses in the Congo regarding large Nile crocodiles. Gene Thomas, missionary in the Likouala since 1955, saw an enormous croc (28 ft, no less), sunning itself on a sand bank near the Motaba River. Gene estimated the croc's size by comparing it to the length of his boat, which was 30 ft. The croc was also estimated to be about 3 ft wide. He even took a shot at it with his high powered rifle.

— Bill Gibbons, cz@onelist.com, July 13, 1998

During a trip around Australia in 1986 I spent some time with two professional crocodile-hunters in the Northern Territory. The first day I met them they told me about the animals they had shot the day before—a giant measuring at least 32 ft. And no, they were not trying to kid me. We went out to see the animal and I brought a tape measure. The croc was big, very big, but "only" 19 ft. My friends were like "I could have sworn...". My point is, that without a photograph with a scale included, even the most well-meaning and honest eyewitness can easily be mistaken.

— Lars Thomas, cz@onelist.com, July 13, 1998

There are 25 extant species of crocodilian (crocodiles, caimans, alligators, gharial, and false gharial) currently recognized by science (if we accept the recent splitting of the dwarf crocodile *Osteolaemus tetraspis* into three separate species as proposed by some taxonomists). However, the cryptozoological archives contain accounts of several remarkable forms that may represent additional, still-undiscovered species or currently unrecognized, giant versions of known ones.

THE MAHAMBA AND LIPATA—MEGA-CROCODILES FROM CENTRAL-WEST AFRICA

During the 1980s, Chicago University biochemist and spare-time field cryptozoologist Roy P. Mackal led two expeditions to the People's Republic of the Congo (formerly the French Congo), seeking an elusive water creature known as the mokele-mbembe, which is said to resemble a modest-sized sauropod dinosaur. Sadly, the team never caught sight of it (though its three-toed footprints were spied), but while there Mackal's team did obtain some native accounts concerning a number of other mysterious, unidentified Congolese beasts. One of these was a gigantic crocodile, known locally as the mahamba.

Science is already aware of the existence in this country's vast

Likouala swamplands of the Nile crocodile *Crocodylus niloticus*, which is Africa's largest known species of modern-day crocodilian, with very old, mature male specimens sometimes exceeding 18 ft. The largest confirmed specimen was a male individual shot near Mwanza, Tanzania, which measured 21 ft. and weighed 1 ton.

According to the Likouala's Bobangi pygmy tribe, however, a type of crocodile far greater in size than even this exceptional Tanzanian specimen exists here, which they refer to as the mahamba. They claim that the mahamba can attain a total length of 50 ft., and insist that it is neither an outsized nor a size-exaggerated version of the Nile crocodile (a species already familiar to them, which they term the nkoli) but is a bona fide, totally separate form in its own right. They also state that it digs lengthy underground tunnels, at the end of which is a cavern where it lives, that it lays eggs, and is carnivorous, even including humans within its diet. Moreover, Mackal's mahamba reports are not the only ones concerning extra-large crocodiles in central-west Africa.

Likely appearance and scale of the mahamba (David Miller/Roy P. Mackal)

Back in 1890, Belgian explorer John R. Werner reported in his book *A Visit to Stanley's Rear-Guard [...] with an Account of the River-Life in the Congo* how on two separate occasions he had personally observed a crocodile whose total length he had estimated at 40-50 ft., using his 42.5-ft.-long A.I.A. steam launch as a scale. On the first encounter, he had spied the monster lying on a low sand-spit, but scared it off when he shot at some ducks nearby. On the second, his vessel had actually run into the creature unawares, which lost no time in diving into deep water in order to escape, and in so doing revealed itself to be longer than the A.I.A.

During the early 1930s, French zoologist Albert Monard investigated reports by Tyipukungu village natives of an outsized mystery crocodilian known locally as the lipata, which is said to inhabit the Kasai and upper Chiumbwe Rivers in northeastern Angola. It apparently spends much of its time hidden in the water, only appearing at the surface around dusk and early morning, and is most active during falls of rain, but hardly ever emerges onto land. Monard employed a strychnine-poisoned pig carcass as bait, hoping to procure a specimen of this formidable reptile, but he did not succeed in doing so.

A ferocious carnivore, the lipata not only devours livestock such as pigs, goats, and cattle, as well as any hapless human that it can seize, but will even attack and consume other crocodiles. As big as or bigger than the largest Nile crocodile (which the Tyipukungu villagers term the ngandu), the lipata is further distinguished from the latter crocodilian by its stockier body, its proportionately larger mouth and wider throat, and its more closely set eyes. In spite of the natives' claims to the contrary, however, Monard concluded that lipata specimens are merely very large individuals of the Nile crocodile. But are they? Veteran cryptozoologist Bernard Heuvelmans opined that the lipata might constitute an unknown extra-large species of dwarf crocodile belonging to the genus *Osteolaemus*.

In 1954, while journeying through the Maika marshes in the Democratic Republic of Congo (formerly Zaire and, earlier still, the Belgian Congo), which borders the People's Republic of the Congo, traveler Guy de la Ruwière reported seeing a very unusual crocodile. It measured approximately 23 ft. in total length, which included an unusually long neck, as revealed when the creature lifted its large head out of the water on several occasions before finally diving beneath the surface, creating a huge wave.

When assessing reports of the mahamba, Mackal recalled the existence in prehistoric times of some truly monstrous crocodilians, including *Phobosuchus*, since renamed *Deinosuchus*. This was an alligator-related form that lived in North America 80-73 million years ago, during the late Cretaceous Period. It has been estimated that the biggest specimens may have grown to a total length of up to 40 ft., and weighed as much as 8.5 tons. Other bygone behemoths of comparable proportions include *Sarcosuchus* (Niger, Cretaceous) and *Rhamphosuchus* (India, Miocene), verifying that colossal crocodilians have indeed existed at one time or another on our planet.

Some mystery beast researchers have speculated that the mahamba may actually be an unconfirmed surviving representative of *Deinosuchus*. So far, however, Mackal has remained content to categorize reports of the Congo's mega-croc as being based upon outsized or size-exaggerated specimens of the Nile crocodile, which is the wisest option. Certainly, over-estimation of an already-large animal's size is a very common error of judgment, from which not even the most experienced of observers are immune, as Danish zoologist Lars Thomas's testimony, quoted at this chapter's beginning, vividly demonstrates.

In any event, as Heuvelmans remarked to Mackal when discussing the mahamba: "Only an examination of a skin or skull can settle the question." And unfortunately, those vital examples of tangible, physical evidence in support of the mahamba's reality are presently conspicuous only by their absence.

THE RAILALOMENA—A HORNED MYSTERY FROM MADAGASCAR

Today, only one crocodilian form—a small dark-colored subspecies of the Nile crocodile—is known to exist on the island of Madagascar. However, possibly until as recently as a few centuries ago, a second, much more unusual form also existed here. This was the Madagascan horned crocodile *Voay robustus*, named after the pair of prominent "horns" extending from its skull's posterior region (these were actually the posterolaterally extended corners of the squamosal bone), and believed to have been at least as long as the Nile crocodile, if not longer.

Reports exist of a mysterious horned reptile known as the railalomena inhabiting the huge swamps that once existed in this island mini-continent. An alternative spelling of its name, railelamena, translates as "father of the crocodiles," and in 1802 French naturalist Pierre Denys de Montfort, writing about crocodile size, stated: "… new and trustworthy accounts say that there are some [crocodiles] in Madagascar which reach 18 m [59 ft]."

Yet even if that measurement is reduced by as much as a third in order to compensate for eyewitness exaggeration, this still leaves a length of around 40 ft., which is twice that of any known species of Africa crocodile alive today. And what about the railalomena's horned state? No confirmed living crocodilian has horns.

Is it possible, therefore, that *V. robustus* lingered here until as recently as two centuries ago, or at least until its highly distinctive image had become firmly embedded in the racial memory of the Madagascan people? It is now known that a number of other spectacular members of the Madagascan megafauna, including various giant lemurs and

Crocodilian Monsters and Mysteries 129

also the great elephant bird *Aepyornis maximus*, survived into historical times (see Chapter 7)—so why not the horned crocodile too?

GIANT CROCODILE-FROGS OF BORNEO

The following information concerning an alleged giant "crocodile-frog" was contained in Czech cryptozoologist Jaroslav Mareš's book *Detektivem v Říši Zvířat* (*A Detective in the Animal Kingdom*) (1995), but it has not appeared in any English-language book before. Hence I am greatly indebted to Czech cryptozoological enthusiast and friend Miroslav ("Mirek") Fišmeister for kindly bringing it to my attention and translating it for me.

Jaroslav Mareš learned about this cryptid from the Seluks, a river-dwelling Dusun tribe in the Malaysian state of Sabah in northern Borneo, while he was leading two expeditions, during 1976 and 1985, in search of a giant specimen of the estuarine crocodile *Crocodylus porosus* nicknamed the Devil's Father.

The Seluks claimed that less than a day's journey by boat from their village, living in the jungle forest on the left bank of Sabah's Segama River, was a very dangerous beast that resembled a huge, 10-ft.-long frog but with the head of a crocodile. Mareš seriously doubted this, but two Seluks from the previous village that he had visited assured him that they had seen it and knew its precise location. Consequently, Mareš visited this area, but throughout the day he and his guides found nothing. Then, on their way back to their boat, they heard what sounded like branches breaking or even stones being crushed—and suddenly, standing at the edge of some bushes, an extraordinary animal was sighted.

Its head and forelimbs were obscured by thick brushwood, but its tailless body resembled that of an enormous frog, and its powerful muscular hind legs were very bulky and fairly long, raised quite high, so that the creature's body pitched forward noticeably. Paunchy and flabby, its body was covered with brown scales spotted with black, which were evenly spaced, forming a symmetrical black pattern over its back. The sounds that they had heard were the crunching of meat and the crushing of bones by the creature's as-yet-unseen jaws.

Mareš's guides were already shaking with fear at encountering this monstrous frog-like beast at such close range and in so unexpected a manner, but more was to come. Abruptly, it rose up on its forelimbs and raised its huge head into view—it was indeed the head of a crocodile!

In total length, the creature measured almost 6.5 ft., and its flanks were paunchy as it stood for a while, as motionless as a statue, towering above the level of the vegetation. But then, it began to move—directly towards Mareš! Having seen all that he desired, Mareš lost no time in retreating to his guides, who had remained at a safe distance behind him, and they all fled back to their camp. Happily, the crocodile-frog didn't pursue them.

Initially, Mareš wondered if this bizarre creature constituted an extraordinary, previously unknown species of tailless crocodile, but upon further reflection he considered it more likely that it was a freak, malformed specimen of the estuarine crocodile that had simply never developed a tail. He ruled out the possibility that it had lost it during some fight with another crocodile; for although this can occur, from his observation of the Bornean crocodile-frog Mareš saw no evidence that it had once possessed one. He also spoke later to Australian crocodile specialist Graham Webb, who informed him that within a single nest of crocodile eggs, several baby crocodiles can hatch that lack tails. The reason for this may be genetic, or it may be epigenetic (due to too high an external temperature during incubation).

Perhaps the giant crocodile-frog was one such unfortunate individual, but which had overcome its handicap and had survived, growing large and powerful by adapting to a life in which it hunted terrestrially. Prey victims such as pigs would be easier for it to catch by gripping their mouths in the usual crocodile manner if it remained concealed by bushes on land, rather than attempting this action while hidden in the water like normal crocodiles, which also use their powerful tail as a weapon to bowl over their victims if necessary—something that the crocodile-frog would not be able to do.

Moreover, it may be sheer coincidence, but the famously tailless Manx cats are also known for their unusually well developed hind legs. Could there be a rare mutant gene in crocodiles whose expression results in this same linked phenotypic effect?

The notion of crocodiles hunting on land is not as unusual as it may initially seem. Although such species have seemingly all died out now,

Reconstruction of the giant Bornean crocodile-frog (Tim Morris)

Crocodilian Monsters and Mysteries

there is conclusive fossil evidence that terrestrial crocodilians used to exist right up until very recently. Furthermore, some were able to walk on two legs, and although all of today's recognised crocodilians live in water, their hind legs remain longer than their front legs, reflecting their terrestrial past. A growing number of different forms have been identified in the fossil record, from parts of Australia and on certain other islands in the Pacific, including those of Vanuatu, Fiji, and New Caledonia, where they survived until the arrival of humans. Members of this group (variously classed as a subfamily or family) are known as mekosuchines.

The last species of mekosuchine to die out, *Mekosuchus inexpectatus*, measuring approximately 6 ft. long, lived on the island of New Caledonia, east of Australia. Some authorities have speculated that it may have persisted here until as recently as 1670 years ago, which is more than 2,000 years after humans first arrived. Indeed, it may well have been hunted into extinction by them, as were several other New Caledonian endemics.

Returning to "crocodile-frogs": further evidence for the existence of tailless crocodiles can be found in a second Czech cryptozoology book, written this time by Vojtìch Sláma, entitled *Hon na Vodní Pøíšery* (*The Hunt For Water Monsters*), and published in 2002. Kindly translated for me once again by Mirek, the relevant information from this book is as follows.

While Sláma was spending a week journeying along the Segama River, Bateig Labi, a well-respected 80-year-old Kadazan (=Dusun) fisherman and patriarch of the village of Bugit Balachon, informed him that on several occasions in the jungles around Segama he had spied a 10-ft.-long creature that the local Kadazans termed a crocodile-frog. He stated that it had a huge crocodilian head with bulging eyes but a large slimy amphibian body.

Sláma freely admitted that it was very inviting to consider the dramatic possibility that the creature seen by Labi was a surviving species of *Mastodonsaurus*—a genus of huge terrestrial temnospondyl amphibian that lived during the mid-late Triassic Period, approximately 200 million years ago. Indeed, with a triangular head measuring up to 4 ft. long, and a total length of 13-20 ft., *Mastodonsaurus* was the largest animal on the entire planet at that time, and was a feared predator.

As Sláma noted, this formidable creature is traditionally reconstructed as resembling a huge frog but with a somewhat crocodilian head, which is clearly a description comparing closely with Borneo's mysterious giant crocodile-frog. However, that reconstruction has lately been challenged, and nowadays palaeontologists prefer one that presents a creature with a longer tail and a more salamander-like or crocodile-like overall form.

In any event, Sláma was much more in favor of Mareš's theory, that the giant crocodile-frog was a malformed tailless crocodile that had overcome its handicap and had survived by preying upon land creatures. In his book, he included a photograph of one of the tailless crocodiles that he had encountered at the Sandakan crocodile farm, and there is little doubt that a freak specimen of this nature might well indeed have inspired native legends of huge crocodile-frogs in the jungles fringing the Segama River.

CROCODILE-JAWED SERPENT DRAGONS OF NEPAL

It might seem scarcely conceivable, but an even more extraordinary variation upon the crocodilian theme than the bizarre crocodile-frogs documented above has also been reported from southern Asia, and as recently as 1980.

That was when Reverend Resham Poudal, an Indian missionary, was leading an entourage through a Himalayan jungle valley in Nepal. They came upon what seemed at first sight to be an enormous log, greenish-brown in color, lying on the ground across their planned path—and then the "log" moved! To the great alarm of everyone present, it proved to be a huge limbless reptile, whose scaly serpentine form blended in so well with the surrounding vegetation that when stationary, it did indeed look exactly like a log or fallen tree trunk.

Its eyewitnesses estimated the creature's total body length to be at least 42 ft., and approximately 6.5 ft. in circumference, but most shocking of all were its jaws. For whereas those of true snakes, even massive ones, are relatively short in relation to their body, this mystery reptile's were extremely long, greatly resembling a crocodile's jaws. And although they were motionless, they were fully open, yielding a gape wide enough for a 6.5-ft.-tall human to stand inside!

The entourage's native Nepalese members informed the Reverend that they considered these "crocodile-snakes" to be dragons, but stated that they were only very occasionally encountered, and even when one was met with, it rarely moved. Instead, it would simply lie impassively with its monstrous jaws agape and wait for unsuspecting prey, usually water buffaloes, to approach, not seeing its enormous yet perfectly camouflaged form until it was too late. For as soon as a buffalo walked within range, the dragon's open jaws would seize it, and from those immensely powerful killing implements, brimming with sharp teeth, there would be no escape. In addition, the natives claimed that its eyes glowed like luminescent lamps at night (a feature also reported for anacondas and other very large snakes), which helped to lure prey.

Nepalese crocodile dragon (William M. Rebsamen)

But if such a remarkable creature as this truly exists, what could it be? Possibly an immense species of snake with unusually large jaws, or perhaps a gigantic legless lizard? Might it even be a unique limbless species of terrestrial crocodilian, highly specialised for this cryptic, motionless lifestyle? Whatever it is, it certainly does not match the appearance of any reptile currently known to science.

Mysterious, still-unidentified forms of crocodilian have not only been reported from various freshwater and terrestrial localities around the globe. As will now be revealed, several equally intriguing and controversial forms have also been reported from the oceans—where no formally recognized species of modern-day crocodilian is known to exist on a permanent basis today.

A MYSTERY GHARIAL FROM AUSTRALIA?

The most distinctive, readily identifiable species of crocodilian alive today is the gharial or gavial *Gavialis gangeticus*, due to its exceedingly long, slender jaws and its very sizeable total length (a few specimens exceeding 20 ft. have been confirmed; only the estuarine crocodile is longer). However, the gharial is confined entirely to the Indian subcontinent. Nothing like it has ever been recorded from Australia—officially...

The huge estuarine or saltwater crocodile (up to 20 ft. long, sometimes slightly more), and the smaller freshwater or Johnston's crocodile *Crocodylus johnstoni* are the only two species of crocodile known to exist in Australia today. However, Australian herpetologist Richard Wells has informed me that when he lived in the Northern Territory several years ago, he received a number of consistent reports from a variety of independent informants, including aboriginals, old ex-crocodile hunters, and fish poachers, suggesting that the tidal parts of its Mary River system was home to a third, very large but dramatically different crocodilian that is apparently unidentified by science.

In one of his communications to me, Wells stated that this mysterious reptile is claimed to be entirely aquatic, and possesses elongated jaws containing numerous exposed teeth, making it reminiscent of the gharial. However, it is said to be even bigger than the latter species, has paddle-like limbs, and is nocturnal. Australian naturalists have tended to dismiss reports of this mysterious creature as misidentifications of known crocodiles or even sawfishes, but those persons who have reported it are familiar with these animals and have vehemently discounted all attempts to identify it with them. Moreover, they were all very frightened by it.

So what could this gharial-lookalike be? Interestingly, crocodiles in zoos that have been fed a low-protein diet sometimes develop exposed teeth that jut out laterally from the jaws. So could the Mary River mystery crocodile simply constitute size-exaggerated freak specimens of one or other of Australia's two known species? Worth noting is that Johnston's crocodile does possess a rather long, superficially gharial-like snout. But this does not explain the paddle-like limbs cited for the Mary River beast (thus rendering the already-improbable possibility of an undiscovered population of bona fide gharials existing here even more improbable).

Intrigued by this cryptid, I contacted Australian reptile palaeontologist Ralph Molnar for his views, and he aired a thought that had occurred to me, too. Namely, that a paddle-limbed crocodilian instantly recalls the thalattosuchians or sea crocodiles.

These constituted a prehistoric group of highly specialized, elongate-bodied, and wholly aquatic marine reptiles related to true crocodilians and superficially similar to them, too. As they lacked the osteoderms of true crocodilians, however, they were smooth-skinned. In addition, they possessed flippers as limbs instead of claw-footed legs, and probably bore a distinctive tail fin. Yet despite being well adapted for their maritime lifestyle, they were a relatively short-lived group, and vanished from the known fossil record over 110 million years ago. But could it be that a lineage of thalattosuchians has somehow survived into the present day, undetected by science and only subtly changed from those ancient times? It seems highly unlikely, and yet it would not be the first time that the existence of a "living fossil" (or Lazarus taxon) has been confirmed.

A much more conservative, alternative option, but still involving prehistoric survival, hinges upon whether the eyewitnesses' claims that the Mary River crocodile's feet are paddled are strictly accurate. Could they simply be more rounded than those of Australia's pair of known crocodile species? For if they are not truly paddled, a very remarkable identity for the Mary River crocodile, as noted by Molnar, becomes available for consideration.

Less than two million years ago, a very gharial-like species known as the Murua crocodile and dubbed *Gavialis papuensis* inhabited the Solomon Sea, just above southeastern New Guinea, and therefore in close proximity to the seas off Australia's Northern Territory. In the past, some authorities were more in favor of classifying this species as a thoracosaur, but Molnar now considers that it may indeed have been a gharial. Certainly, its elongated snout was very like the latter's, as was its piscivorous diet. Based upon what is known of its morphology from fossil evidence, the Murua crocodile would correspond very closely indeed with the Mary River's mystery reptile, though it was smaller than the latter cryptid is said to be (but as noted earlier here, this greater size may simply be a product of exaggeration or poor estimation on the part of eyewitnesses).

SEA SERPENTS OR SEA CROCODILES?

Many different types of so-called sea serpent have been reported and categorized over the years, a few of which exhibit some decidedly crocodilian attributes, as shown by the following examples.

One morning in May 1901, Charles Seibert was one of several passengers aboard the steamer *Grangense* who saw a remarkable but still-unidentified creature frolicking in the sea nearby at the mouth of the Amazon River in Brazil during their journey from New York to Belém, Brazil. According to their descriptions, it had a crocodile-shaped head whose jaws contained rows of 4-6-in.-long teeth, a short neck, and a long greyish-brown body. As they watched, it disported in a seemingly playful manner, twisting around in horizontal circles before finally diving out of sight beneath the waves.

Reconstruction of the *Grangense* sea serpent (Tim Morris)

A similar if equally mystifying sea beast had been reported back on July 30, 1877. This was when Captain W.H. Nelson and a helmsman aboard a ship named the *Sacramento*, traveling in the mid-Atlantic during a journey from New York to Melbourne, Australia, observed an astonishing creature resting at the surface of the sea with its head raised about 3 ft. above it. They considered its head to be very alligator-like in appearance, but about 10 ft. behind it was a pair of flippers. Reddish-brown in color and estimated by its two eyewitnesses to measure approximately 50-60 ft. long, it was very elongate in shape. Eventually, while still being watched, the creature began moving away from the ship, lowering its head after having attained a distance of 30-40 ft. from it.

For over a century, reports have emerged from New Zealand of huge sea monsters resembling flipper-limbed lizards but with crocodile-like heads encountered by terrified sea-goers in the waters around this dual-island country. One such animal, with noticeably big eyes, was spied about 18.5 miles off Lyttleton during April 1971 by the crew of the *Kompira Maru*, who noticed its flippers when it dived underwater. Another sighting, featuring what its terrified eyewitnesses termed "a giant lizard," 13-16.5 ft. long and green in color, occurred in 1990. This was when two young women sunbathing by a lagoon close to Taupo saw the creature in question swimming in the shallows. It even attempted (unsuccessfully) to catch a bird in its jaws, the front portion of its body emerging from the water during this attempt, but then it submerged again and swam into the depths.

In 1993, Earl Rigney from Canterbury realized, after viewing it through a telescope, that what he had initially assumed to be a whale far out to sea was actually a colossal crocodile, roughly 30 ft. long (i.e. almost a third longer than the largest estuarine crocodiles), breaching at the water surface. And during September of that same year, while fishing off the Cook Islands (lying approximately 2,000 miles northeast of New Zealand), a vicar and his son allegedly spied a creature resembling a huge lizard, bigger than a whale, surface near to their vessel. This so alarmed them that they abandoned their fishing and sailed away with all speed!

So how do we explain such reports as these? The estuarine crocodile has sometimes been found far out to sea and is a good swimmer. Could it be, therefore, that these maritime mystery beasts are nothing more than misidentified specimens of the latter species, flippers notwithstanding? Or is a surviving, scientifically reclusive species of thalattosuchian implicated, whose continued evolution over more than 110 million years has wrought various external morphological changes so that it no longer corresponds precisely with its fossilized antecedents? Or perhaps, more conservatively, a species is responsible that has evolved only fairly recently from true crocodilians, within just the past couple of million years or so, to yield a crocodilian more adapted for an exclusively aquatic existence than all other modern-day species but not dramatically different in overall body form from them?

Some cryptozoologists have rejected any crocodilian link at all, variously proposing instead that such beasts are surviving mosasaurs or surviving pliosaurs. Mosasaurs were enormous flipper-limbed sea-lizards with huge jaws, related to today's monitors (varanids), and up to 57 ft. long; whereas pliosaurs were short-necked, massive-jawed plesiosaurs, up to 49 ft. long. All of these reptiles, however, were officially extinct by the close of the Cretaceous Period 65 million years ago.

Without a specimen to examine, we can only speculate concerning identities, but if the reports presented here are genuine and their descriptions accurate, it does seem likely that a marine crocodilian form decidedly different from any species currently accepted by science is indeed out there somewhere in the vast oceans, eluding all but the sharpest-eyed and most fortunate of observers.

THE U-28 SEA SERPENT—A WAR-TIME CONTROVERSY

No review of mystery marine crocodiles can be considered complete without examining the most (in)famous case on file—that of the *U-28* sea serpent.

On July 30, 1915, during World War I, the British steamer *Iberian* was torpedoed off Fastnet Rock, Ireland, by the *U-28*, a German submarine. This much is fact. What remains a subject for considerable conjecture in cryptozoological circles, conversely, is what *may* have happened immediately afterwards. For according to Georg Günther Freiherr von Forstner, captain of the *U-28*, a violent underwater explosion, which occurred less than a minute after the doomed steamer *Iberian* had sunk, blasted 65.5-98.5 ft. out of the water not only some wreckage from it but also a huge, living sea monster!

Terming it "the underwater crocodile," the captain (and crew) could observe it in its entirety for the 10-15 seconds that it was above the water surface before falling back down into the sea again. He likened its shape to a crocodile's and stated that its head was long and tapering (although some versions of his account report that it was its tail that was long and tapered to a point), and its four limbs had powerful webbed feet. However, he estimated its total length at a huge 65.5 ft.

If genuine, this would be a remarkable report—but *is* it genuine? Investigators have uncovered a disturbing number of problems. Why, for instance, did the captain not mention such an extraordinary incident until autumn 1933 (which just so happens to be only a short time after the Loch Ness monster had first made news headlines worldwide), when it featured in a German newspaper report? And for someone we would expect to be able to estimate sea depth with a marked degree of accuracy, he showed a singular lack of such ability when discussing this case. For he claimed that the *Iberian* must have sunk to a depth of roughly 500 fathoms prior to the explosion—and yet as this explosion had occurred a mere 25 seconds or so after the ship had been torpedoed, the speed of submergence would have needed to be a highly unlikely 90 miles per hour! And why did he refrain from revealing the incident's

The *U-28* sea serpent blasted into the air (public domain)

precise location until as late as 1942?

Most baffling and worrying of all, however, is that not the slightest mention of any sea monster being blasted to the surface was included in any local Irish newspaper's coverage of the sinking of the *Iberian*, even though one of them published an account of a much less spectacular sea serpent sighting elsewhere just a week later (thereby demonstrating that it was not averse to covering such stories). Nor has any surviving crewmember of the *U-28* (which was itself sunk on September 2, 1917) ever spoken of this bizarre incident.

Little wonder that many cryptozoologists nowadays deem the curious case of the *U-28* sea serpent to be a total invention—its credibility blown out of the water in a way that the fictitious beast itself never was.

One thing *is* definite, though. Regardless of their respective zoological identities, if any of the mystery crocodilians reviewed here are indeed genuine, they are likely to be among the most spectacular reptilian cryptids ever recorded on Planet Earth.

Chapter 11:
MEGA-BEAVERS AND MICRO-SQUIRRELS—
TWO EXTREMES IN CRYPTO-RODENTOLOGY

Many, many suns in the past, ere the wigwams of our tribe stood here, a great lake rippled wide and long across the land. In its waters a giant beaver sported, and ravaged all the countryside. Mighty Hobomuck [a benevolent spirit giant], wroth, vowed that the wicked one should die. With an oak cudgel he struck across the beaver's neck—just there, O Netop, in the hollow between head and shoulders. The fearful creature sank gasping to the bed of the lake and his carcass turned to stone.

> — K.M. Abbott, *Old Paths and Legends of the New England Border*

I have seen some extraordinary sights at one time and another, but the flight of the flying mice I shall remember until my dying day...They swooped and drifted through the tumbling clouds of smoke with all the assurance and skill of hawking swallows, twisting and banking with incredible skill and apparently little or no movement of the body.

> — Gerald Durrell, *The Bafut Beagles*

Mystery rodents come in many shapes, colours, and sizes—but few occupy more extreme positions on the scale of magnitude than the two examples presented here, the second of which has never before been documented in any cryptozoological publication.

THE GIANT BEAVER—A LEGEND REBORN?

Approximately 3 ft. long and generally weighing 33-77 lb., the American beaver *Castor canadensis* is second in size only to the capybara among present-day New World rodents. According to Amerindian traditions, however, there was once a gigantic form of beaver existing in the U.S. and Canada that exceeded even the capybara in stature.

The Malecite are an Algonquian-speaking Native American people indigenous to the St. John River valley, crossing the borders of New Brunswick and Quebec in southeastern Canada and Maine in the northeastern U.S. Their time-honored orally preserved lore contains a detailed legend of how Gluskap, their mythical culture hero and transformer, angrily pursued a giant beaver for failing to show due respect to his latest creation, man, and for building huge dams that blocked the river. However, the giant beaver escaped his clutches, fleeing far away.

Is this just an entertaining myth, or is it an example of a

Scale drawing of capybara, top left; American beaver, top right; and extinct giant beaver *Castoroides*, bottom (Tim Morris)

preserved racial memory of early encounters with a member of North America's now-extinct megafauna? Or could it even be a bona fide case of prehistoric survival?

For as will be seen, giant beavers are not confined to legends and folklore. Several thought-provoking reports of mysterious aquatic beasts very reminiscent of such super-sized rodents can also be found within the bulging archives of cryptozoology.

First of all, however, let's consider some more native myths, demonstrating that such stories are by no means confined to eastern North America's Malecite culture. In British Columbia, southwestern Canada, for instance, there is a longstanding Salishan tradition of a huge beaver-like beast known locally as the slal'i'kum inhabiting Cultus Lake. A legend from the culture of the Pocumtuck tribe inhabiting the region around Deerfield in Massachusetts tells of how Lake Hitchcock—a large lake in the Connecticut River Valley that dates back to the Pleistocene epoch—harbored a giant beaver that sometimes came ashore and attacked people, until one of their bravest hunters killed it. They also claim that several lakes here were created as a result of dams constructed by gargantuan beavers.

The lore of the Tlingit people living around Sitka on southeastern Alaska's Baranof Island contains the story of how a gigantic beaver-like beast once devastated an entire village. And the Montagnais-Naskapi people, an Algonquin tribe inhabiting Labrador, allegedly named a major river there Mishtamishku-shipu (Giant Beaver River) as a direct reference to a pair of enormous beavers, said to be larger than seals, that were killed there long ago in order to prevent them from breeding.

Moreover, claims concerning giant beavers have been made in much more recent times too. During the 19th century, reports of a mystifying brown-furred water beast likened to a huge beaver or immense otter

Mega-Beavers and Micro-Squirrels 143

emerged from Fowler Lake in Saskatchewan, though, sadly, the beast itself did not do likewise. Accounts of what some investigators deem likely to have been an extra-large beaver-like creature are also on file from Utah's Bear Lake and Utah Lake during that same time period. And sightings of creatures recalling the slal'i'kum have been reported in British Columbia's Cultus Lake as recently as the 1990s. Moreover, some fascinating reports from as recently as the late 20th century and present 21st century were collected and published by American cryptozoological chronicler Michael Newton in his *Encyclopedia of Cryptozoology* (2005).

In March 1993, Tom Greene, horticultural superintendent for Moline, Illinois, claimed that a 5-ft.-long beaver weighing more than 75 lb. and extremely strong was gnawing down trees at the Marquis Harbor Yacht Club. It had already sprung and escaped from two traps set to snare it.

During 2000, a Ms. J. Greenwald supposedly spied a horse-sized beaver close to Bullfrog Marina, on Lake Powell, a sizeable body of freshwater that spans the southern Utah-northern Arizona border in the southwestern U.S. Two years later, she finally reported her sighting formally to the National Institute of Discovery Science in Las Vegas, Nevada. In her report, she stated that the creature was brownish-black, weighed an estimated 700-800 lb., was unbelievably big, and could certainly kill a person if it so chose. Yet even assuming that she had undoubtedly over-estimated its size, if her claim is true then it still must have been a seriously big beaver!

Interestingly, in August 2002 an online account posted by someone identified only via the user name Staci told of how her family had encountered a beaver "the size of a horse" at Lake Powell. In view of the very same comparison of stature (horse-sized) and location (Lake Powell) appearing in both reports, could Staci have been one and the same as J. Greenwald, I wonder, reporting an anonymous version of her sighting before going public with it during that same year?

Even more recently, in 2007, I was contacted on several occasions by Canadian cryptozoologist John Warms who is actively investigating reports of alleged giant beavers, following his own sighting in 2006 of what he believes to have been one such creature. Based in Manitoba, he had first learned of their apparent existence back in 2003, after speaking with residents of a Cree Nation reservation in northern Manitoba. Since then, he has been in touch with a number of other eyewitnesses, all describing similar creatures but this time without realizing that extra-large beavers were unusual.

After hearing about bygone observations of huge beavers in Manitoba's Assiniboine River, Warms decided to spend a night in his Jeep beside this river during April 2006, when it was in flood stage, and here, in his own words, is what he saw:

> Just before dark, not expecting to see anything, I saw this large head and what I took to be its back and tail behind it, moving along with the current. As I bent down to have a better look through the branches, it immediately plunged and slapped the water like an ordinary beaver. I was stunned to think that I had perhaps seen the "extinct" giant beaver, and had to convince myself ever since that what else could it be? The head was the size of a basket ball, and the whole body was seven or eight feet long. I have seen hundreds of ordinary beaver, and there was just no comparison.

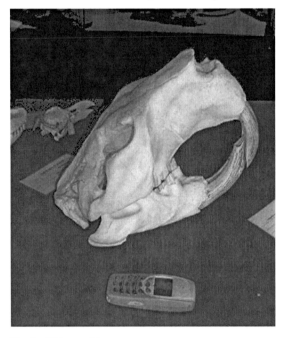

Skull of *Castoroides ohioensis* (Markus Bühler)

Warms informed me that he has also found some very large underwater tunnels, 3 ft. in diameter, that he considers may have been constructed by giant beavers, particularly in southern Manitoba, and he believes that they occasionally build lodges and dams too, like America's (and Europe's) smaller, known species of beaver. Warms is hoping to elicit funding to conduct extensive research, and during late December 2009 his ongoing quest for evidence of these creatures' existence led him to Utah and Arizona, as featured in various newspaper articles.

Needless to say, the prospect of giant beavers may seem highly improbable, at least on first sight. In reality, however, nothing could be further from the truth, because there is a very significant, fully confirmed precedent. Until at least as recently as the end of the last Ice Age, 12,000 years ago, during the late Pleistocene epoch, when the first human colonists had already arrived here, North America was home to no less than two separate species of bona fide giant beaver.

The better known of these veritable behemoths of beaverkind, and as big as the American black bear *Ursus americanus*, was the northern giant beaver *Castoroides ohioensis*. Fossil evidence reveals that it was up to 8 ft. long, with 6-in.-long incisors, and weighed 132-220 lb. Moreover, the second species, the southern giant beaver, *C. leiseyorum*, was slightly bigger, but its remains are currently known only from South Carolina and Florida, whereas *C. ohioensis* was widely distributed in the midwestern U.S. south of the Great Lakes and also in Canada north of them, as well as in Alaska. Currently, there is no official confirmation that *Castoroides* constructed lodges and dams, but some experts deem this to be likely.

Could it be, therefore, that before its extinction, the northern giant beaver's dramatic form and presence had exerted a sufficiently profound impact upon North America's early human colonists for its memory to have been preserved down through countless generations and incorporated into their orally-preserved folklore, and which also traveled westward with them as they traversed the continent from east to west? Or is it even conceivable that this dramatic species actually lingered on beyond the last Ice Age into historical times, and perhaps, just perhaps, even into the present day in certain remote localities within its prehistoric range?

The last word on this thought-provoking subject comes from a paper published in spring 1972 in the highly reputable periodical *Ethnohistory*, and which first alerted me to the fascinating presence of giant beavers in cryptozoology. Written by Jane C. Beck, it is entitled "The Giant Beaver: A Prehistoric Memory?" and evaluates the frequent presence (with only minor variations) in Native American tradition of the basic folktale concerning a deity pursuing a giant beaver. In her conclusion, Beck offers the following hypothesis to explain the unexpected abundance and persistence of what would ordinarily be a very local, obscure legend:

> Thus it seems that all evidence points to the giant beaver tale being a folk memory of a prehistoric creature. Actual proof is not yet possible, but it is important to look to the future and suggest a road that the archaeologist and historian might well follow.

This suggestion has already borne fruit. Just over 20 years after Beck's paper was published, the second, southern species of giant beaver was formally described and named, and continued findings of fossil remains has expanded the northern species' known prehistoric range very considerably. In short, new revelations concerning North America's onetime giant beavers are still occurring. So who knows, perhaps in the future, further tangible, physical discoveries will be made that will verify post-Pleistocene persistence of these mega-rodents after all.

SCHOMBURGK'S MISSING MICRO-SQUIRRELS

It's often been said that good things come in small packages, and this is certainly true of the following case, in which the cryptid in question may be minuscule but is definitely no less memorable for that.

The German explorer/naturalist/film-maker Hans Schomburgk (1880-1967) earned a lasting, well-deserved place in zoological and cryptozoological history by rediscovering on June 13, 1911, the pygmy hippopotamus *Choeropsis liberiensis*, alive and well in Liberia—where it had long been known to the native people as the nigbwe, yet had been ignored by scientists. Indeed, until then this enigmatic species had generally been discounted as nothing more than a freak of nature, consisting merely of dwarf, stunted specimens of the larger, common hippopotamus *Hippopotamus amphibius*, or even as a juvenile form of the latter—despite American zoologist Samuel G. Morton having officially named and described it as a valid, second species of hippo more than half a century earlier in 1849. Following its rediscovery by Schomburgk, however, studies confirmed its status as a valid and very distinct species in its own right. (For full details concerning the pygmy hippo's controversial history, see my book *The Encyclopaedia of New and Rediscovered Animals*, 2012.)

Yet whereas Schomburgk became famous for refinding this erstwhile cryptid, it was by no means the only creature of cryptozoology that he had heard about during his numerous African expeditions. In his book *Zelte in Afrika* (*Tents in Africa*), which was published in 1957 and looked back on his explorations and researches from the previous six decades, he documented a number of very interesting mystery beasts, some of which were new to me when they were brought to my attention in July 2012 by German cryptozoological investigator Markus Bühler.

Although I'd heard of Schomburgk's book, and had read a little concerning his cryptozoological accounts in other publications, I'd never seen a copy of it. Nor had I received any excerpts or summaries from it, until Markus kindly provided me with the following information:

Mega-Beavers and Micro-Squirrels 147

There are so many cryptids in Schomburgk's "Zelte in Afrika," including some he has seen himself, which were nowhere else documented...

He wrote for example he once encountered several small mammals which looked like tiny squirrels, and which were extremely tame. They were so cute he didn't want to kill one for his collection, and later he learned these animals were never seen before (and never again)...

There were also some stories which he heard from natives, but without further information, about a giant hyaena, a "jungle tiger" and I think also a mokele-mbembe-like animal. If I remember correctly he also brought a sculpture of some kind of water monster back to Germany, which was in the archives of the ethnological museum at Hamburg, but later given back (but there is still a copy I think). It's already been some time ago since I read the book the last time, but there are really a whole lot of great stories, and a lot of information about water elephants.

As I was so intrigued by the tantalising snippets above, Markus checked through his copy of *Zelte in Afrika* for further details, and this is what he found:

Luckily I have put a bookmark at the pages which mention some cryptids.

So here are some animals which are sadly only mentioned: the lion-tiger (not jungle tiger as I thought) of Senegal, the giant hyaena, the water-leopard [of] Mafue [near Zambia's Lukanga Swamp] which is depicted in the sculpture he discovered.

He also heard reports about chimpekwe [aka chipekwe] at Lake Bangweulu [in Zambia].

He mentions also a creature called koo-be-eng, some kind of giant snake-like reptile with a horn on its head, which lives in the water and eats big animals. He wrote that there is a depiction of such a beast in a cave at Brackfontein [in South Africa].

There is also kou-teign-koo-rou, lord of the water, which was said to be bigger and stronger than a hippo and also living in swamps. Bushmen caught it with very strong traps, and he mentions also depictions.

He also heard tales about the animal tu from the upper part of Morfi River [in Liberia], which was said to be as big as a goat, with teeth of a dog, black fur, and to be quite vicious.

And a giant ape from Vey-Land which was like a huge greyish yellowish chimp with long fur, which kills or mutilates humans.

Although all of these beasts sounded very interesting (I was already aware of the tu, chipekwe, and water leopard, but not the others), the one that fascinated me most was the Liberian micro-squirrel. For although Schomburgk had stated that these delightful little creatures had never been seen again since his encounter with them, they instantly reminded me of an obscure but formally-described and definitely still-surviving species that I had read about somewhere else. So I asked Markus if he could provide me with a loose translation into English from Schomburgk's book of the entire passage describing them, and here it is: (Thanks, Markus!)

> During my search for *Choeropsis liberiensis* I found a bush which was full of lovely small animals. Full of beans, they were. They were grey-brown coloured, reminiscent of tiny squirrels. I put my hand in the bush. The tiny cute beasts whizzed on my palm, stood upright like small rabbits and jumped from finger to finger. Noticeable was the long tail, with feather-like erected hairs. When I tried to hold one on it, it broke off. I repeated this experiment, but always with the same result, so brittle was this, with a thin skin-covered chain of tiniest vertebrae. Surely they were some kind of pygmy mouse. As I was on a hunt, I had no vessel with me, in which I could have caught one of the beasties alive. Only the spirit jar, which one of my native helpers carried. I let him uncap it, to include at least one or two of these beings in my collection. But at this moment the tiny mice looked at me with their big wide eyes in such a clever trusting way, I had not the heart to exploit their trust in the giant who was so mysterious to them. It doesn't have to be today, I thought. I will come back tomorrow or the day after tomorrow and incorporate the lovely bush pygmy mouse into my menagerie. But never again, neither back then nor later, have I or any other explorer found any sign of this odd little animal. And what grieves me most: since then, I have learned that I'd had an animal in my hand that had never been seen in my homeland and which was unknown to science. I have enriched zoology with some discoveries: the *Bubalus schomburgki* (a type of Liberian buffalo), the already mentioned shell *Mutela hargeri schomburgki*, five species of earthworm, from which two have my name as well. However, my lively little mouse belongs to those animals that exist yet which have still not been secured for science.

After reading this account, conversely, I was even more convinced that I knew the identity of Schomburgk's mystifying little mammal, because many years earlier I had read about another famous naturalist

Mega-Beavers and Micro-Squirrels

not only seeking but successfully encountering what seemed to be an identical mini-beast elsewhere in West Africa. The book in which this search appeared was *The Bafut Beagles* (1954), recounting the author's many adventures during a private expedition to Cameroon in 1949, collecting wild animals to sell to zoos. And the name of that author? None other than Gerald Durrell.

One creature that Durrell was particularly anxious to find and collect while in Cameroon belonged to a fascinating but scarcely known genus of rodents—*Idiurus*, the flying mice (a somewhat unfortunate name, as they are not mice, and they glide rather than fly!). And the species of *Idiurus* that he was seeking was *I. kivuensis* (which, as will be seen, has a somewhat convoluted taxonomic history).

Idiurus is one of three genera of squirrel-like rodents belonging to the taxonomic family Anomaluridae. Exclusively African, its members are known collectively as anomalures ("strange tails") or, more colloquially, as scaly-tails, because they are characterized by two rows of overlapping keeled scales on the underside of their tail near the base. Moreover, all but one species also possess a pair of gliding membranes, linking their front and hind limbs on each side of their body.

A flying mouse *Idiurus* (public domain)

Consequently, scaly-tails superficially resemble true flying squirrels or petauristids, but their cranial structure is very different, so they are only distantly related to those latter rodents. Instead, their outward similarity is due to convergent evolution (so too is that of the even more distantly related flying phalangers or petaurids, a group of squirrel-like gliding marsupials native to Australia and New Guinea).

The smallest scaly-tails are the aforementioned flying mice, also called pygmy scaly-tails, belonging to the genus *Idiurus*. Today, only two species are recognised—*I. macrotis*, the long-eared flying mouse (formally described in 1898, and native to western and central Africa, ranging from Sierra Leone to the Democratic Republic of Congo); and *I. zenkeri*, Zenker's flying mouse (formally described in 1894, and native to central and central-west Africa, ranging from Cameroon to the Democratic Republic of Congo).

At the time of Durrell's expedition, a third species was also recognised—*I. kivuensis*, the Kivu flying mouse. When this was originally described in 1917 by Swedish mammalogist Einar Lönnberg, he had categorized it as a subspecies of *I. zenkeri*, but in 1946 mammalogist Robert W. Hayman from London's Natural History Museum elevated it to the level of species. Thus it remained until 1963, when it was reclassified by Belgian mammalogist Walter N. Verheyen as a subspecies of *I. macrotis*, a status that it has retained ever since.

So the *Idiurus* that Durrell encountered in Cameroon during his 1949 expedition there—and documented in a charmingly entitled chapter "The Forest of Flying Mice" in his book—is nowadays deemed to be a subspecies of *Idiurus macrotis*, the long-eared flying mouse.

In his book, Durrell described *Idiurus* as being very similar to the common house mouse *Mus musculus* in both size and general shape, but the most instantly noticeable feature was its very long tail, as it was nearly twice the length of the creature's body and fringed on both sides with long wavy hairs so that it was very feather-like in appearance. Its large, domed head had prominent black eyes, small pointed ears, and a pair of huge bright-orange incisors that protruded in a curve from its mouth. Most interesting of all, however, was its gliding membrane, a long fine flap of skin stretching along each side of its body, attached to each ankle and to a long cartilaginous shaft projecting from each of its forelimbs, just behind the elbow. This membrane remained folded up along each flank when *Idiurus* was resting, but when it jumped into the air and stretched out its tiny limbs the membrane was extended in taut fashion, so that it resembled and functioned like a glider's wings.

Later on, by forcing smoke inside a tree containing more *Idiurus* specimens in the hope of flushing out and capturing some without harming them, Durrell was able to see these tiny creatures become airborne, and what a remarkable sight it was. Leaping into the air and immediately transforming into mini-gliders, the *Idiurus* showed consummate skill in directing their aerial movements. One individual was observed jumping off a tree at a height of 30 ft. or so, and gliding across a clearing in such a straight, well-controlled glide that when it landed on another tree about 150 ft. away, it had lost little if any height, having thus performed an almost perfectly horizontal glide. There was not even any breeze that could have assisted in providing buoyancy or lift, so the extraordinarily adept and intricate airborne feats of the *Idiurus* were entirely achieved by their own remarkable abilities and efforts.

As its common name suggests, a flying mouse does indeed look very murine in general form, because when at rest its gliding membranes are folded up tightly and thus are not readily noticeable (which would explain why Schomburgk never mentioned them). What *is* noticeable, conversely, as mentioned by Durrell, is its very lengthy, plumed tail, which is so fragile that it could certainly be snapped off if not handled with great care.

Moreover, whereas most scaly-tail species are solitary, those of *Idiurus* are colonial; indeed, the two *Idiurus* species have even been found associating together. In 1940, American cryptozoologist and animal collector Ivan T. Sanderson recorded finding approximately 100 individuals of both *Idiurus* species living together in the same tree during his participation in the Percy Sladen Expedition to the Mamfe Division of Cameroon.

Due to their primarily nocturnal lifestyle and highly elusive nature, however, even today the *Idiurus* scaly-tails remain virtually unknown not only to science but even to local hunters—as Durrell discovered during his *Idiurus* searches in Cameroon.

In conclusion, as can clearly be seen from these accounts, *I. macrotis*, the long-eared flying mouse species of scaly-tail, whose zoogeographical range includes Liberia, so greatly resembles Schomburgk's description of his mysterious Liberian micro-squirrel that there can surely be little doubt that the two mammals are indeed one and the same species. This opinion was also supported by Markus after I'd informed him of my thoughts concerning this case.

The mlularuka or Tanzanian flying jackal (public domain)

Another longstanding if hitherto little-known cryptozoological mystery is duly solved, it would seem. Yet many others documented by Schomburgk remain unresolved—at least for now!

Interestingly, this is not the first time that a scaly-tail has been at the core of a cryptozoological mystery. This was ably demonstrated by the remarkable case of Tanzania's flying jackal, which I originally investigated in my book *Extraordinary Animals Worldwide* (1991) and returned to in its updated edition, *Extraordinary Animals Revisited* (2007).

As reported by Captain William Hichens in the December 1937 issue of a monthly magazine entitled *Discovery*, kraalsmen in what was then Tanganyika (now Tanzania) affirmed that an amazing beast known to them as the mlularuka or flying jackal, yet wholly unknown to science at that time, would often raid their mango trees and pomegranates during its flights at dusk, giving voice to loud cries while on the wing. Not surprisingly, their reports were totally disbelieved and dismissed as arrant fantasy—until the "flying jackal" was discovered!

As I learned in a letter of May 8, 1990 from Curatorial Associate Maria E. Rutzmoser of Harvard University's Agassiz Museum, in 1926 zoologist Arthur Loveridge was in Tanganyika, collecting specimens for the museum, and at Vituri he succeeded in tracking down and collecting no less than three specimens of the mlularuka, but it was not a flying jackal. Instead, it was a large (2.5-ft.-long) form of scaly-tail.

Although well known in West and Central Africa at that time, their existence in East Africa had not previously been suspected. Consequently, the mlularuka was initially believed to be a new species but was later shown to be a subspecies of an already wide-ranging species variously called Lord Derby's or Fraser's scaly-tail *Anomalurus derbianus*. Hence it is now referred to as *A. d. orientalis*—a mythical flying jackal that was ultimately unmasked as an aerial squirrel-impersonator!

Sometimes, as exemplified by the resolution of the mlularuka riddle, a mystery is simply the answer to a question that we have yet to formulate.

Chapter 12:
FLOWER-GENERATED BIRDS AND TURTLE-HEADED EELS—EXTRACTING THE ORDINARY FROM THE EXTRAORDINARY IN CRYPTOZOOLOGY

The lands of the north are perpetually snow-bound, but here [in mainland Asia] it was springtime and the snows had melted sufficiently to allow for the growth of mosses, plants and flowers and the breeding of small animals and birds. Here again, we observed a species which combines the animal and vegetable worlds. This is a variety of grass with flowers which take the forms of small brightly coloured birds.

— Una Woodruff, *Inventorum Natura*

Judging from what I see now, it's an eel-like fish. It must be an ancestral or primitive fish. It had fins. But if it's a fish, where are the ribs? It's definitely not a mammal.

— University of Philippines zoologist Perry Ong on the Masbate sea monster, *South China Morning Post*, February 25, 1997

Not all cryptozoological conundra remain unsolved. Some of the most remarkable mystery beasts have ultimately turned out to be surprisingly mundane when formally investigated. This book's final chapter surveys a diverse selection of fascinating examples, drawn forth from the darkest recesses of unnatural history and duly exposed in all their embarrassing ordinariness for all of us to see.

THE FLYING TURTLES OF HENAN

On June 15, 2012, I received via a clickable link placed upon my Facebook page's wall by fortean colleague Robert Schneck an extraordinary illustration, reproduced here. It portrays some supposed flying turtles, native to Henan in China. This bizarre image appeared in German Jesuit scholar Athanasius Kircher's tome *China Illustrata* (1667). Or, to give it its full title, *China monumentis, qua sacris qua profanis, nec non variis naturae and artis spectaculis, aliarumque rerum memorabilium argumentis illustrata*.

As I could think of few creatures less likely to have acquired a mastery of the air than a turtle, I was naturally perplexed and piqued by curiosity in equal measures. *China Illustrata* had originally been published in Latin, but browsing online I soon discovered an English translation of it, and sure enough, there inside I found not

The flying turtles illustration from *China Illustrata*, 1667 (public domain)

only the illustration but also an explanation of it by Kircher, which read as follows:

> The *Chinese Flora* says that in the kingdom of Honan [=Henan, nowadays a province in central China] are found turtles which are green or blue, and that there are also some with wings on their feet, who in this way they compensate for the slow progress they can make on foot. I, however, could not easily believe that these swimming creatures have wings, for it seems to violate the primary nature of a turtle. Rather, turtles give off a sticky liquid around their feet, as the drawing shows, and in time this becomes cartilaginous and resembles a limb which flaps around as they move. This is not used for flying, so when the matter is examined, it turns out to be different than is commonly believed.

The *Chinese Flora* referred to by Kircher was *Flora Sinensis* (1656), authored by Polish Jesuit missionary Michael Boym. It was one of the first European books ever to have been written about China's natural history; despite its title, it included information concerning a number of animals as well as plants.

As for China's flying turtles, here was a situation where one mystery appeared to have been solved only by the citing of another one, because the notion of turtles' feet secreting a sticky substance that hardens to yield flapping quasi-wings was certainly new to me. Happily, however, at the very same time that I was pondering this riddle, a second Facebook-mediated message was en route to me from Robert Schneck that succinctly demystified this entire matter.

In this latter message, Robert informed me that, according to his own investigations, Kircher's documentation of a supposed Chinese belief in flying turtles was an error caused by mistranslation of Chinese sources. What these had actually referred to were turtles with moss, algae, or weeds growing upon their limbs, but unfortunately this had been mistranslated by Kircher (or by earlier non-Chinese works that he had consulted), yielding turtles with wings on their limbs instead.

In short, a seemingly impossible beast had been unmasked as entirely plausible after all, albeit very different in form from how it had originally been described. The flying turtles of Henan may never have existed, but at least they inspired a delightful illustration, one that also serves as a very evocative reminder of the perils of mistranslation—or, how Chinese whispers can engender Chinese flying turtles!

FLOWER-GENERATED BIRDS?

Within my library at home are quite a few delightful works of what I refer to as pseudozoology. Most of these are large, lavishly illustrated books purporting to be republished tomes of arcane natural history, but which upon reading are swiftly recognized as adroitly constructed fiction penned with tongue very firmly in cheek. An excellent example of this highly specialized genre is a truly spectacular tome entitled *Inventorum Natura: The Expedition Journal of Pliny the Elder* (1979), compiled and exquisitely illustrated by celebrated fantasy artist-author Una Woodruff.

The premise behind this very elaborate and skillfully prepared volume is that it is a painstaking reconstruction of a supposedly long-lost work written in Latin by the real-life Roman author-naturalist Pliny the Elder (who died during the eruption of Mount Vesuvius in

79 AD), describing the astonishing fauna and flora that he allegedly observed during a purported three-year expedition to distant lands, an incomplete version of which Woodruff happened to rediscover. The creatures documented in it include many famous legendary beasts, including the basilisk, manticore, unicorn, griffin, vegetable lamb, Chinese hua fish, merman, and dragons. However, it also included a few examples that I had never read anything about elsewhere, so I wondered whether these might have been specially created for this book.

One of them was the bird plant, which according to *Inventorum Natura* was a variety of grass native to mainland Asia whose flowers transformed into small brightly colored birds. The accompanying illustration revealed how these flowers gradually metamorphosed into birds, which eventually broke free of the plant's flower stems to become independent, free-flying entities comparable in every way externally to genuine egg-hatched birds but remaining wholly botanical internally.

Other bird-engendering plants featured in an equally sumptuous work of pseudobotany by Woodruff, entitled *Amarant: The Flora and Fauna of Atlantis by a Lady Botanist* (1981). One was portrayed on its front cover, too.

As I had never seen any mention of bird-generating plants in genuine bestiaries from medieval or earlier times, however, I had assumed that Woodruff's versions were entirely the product of her own wonderfully fertile imagination. Then, while perusing Athanasius Kircher's aforementioned tome *China Illustrata* in search of explanations for his bizarre picture of flying turtles, I was startled yet delighted to discover the following highly illuminating section regarding bird plants:

In Suchuen [sic—Sichuan] Province there is said to be a little bird which is born from the flower called Tunchon, and so the Chinese call it Tunchonfung. The Chinese say that this measures its life by the life of the flower, and that flower and bird die at the same time. The bird has a variety of colours. When flying and beating its wings, the bird looks like a beautiful flower flying across the heavens. Whether an animal, bird, or insect could really be produced from a plant is doubtful. We have denied this in Book Twelve of our *Subterranean World*. It is not possible for the vegetable level of nature to progress to the sentient, since it is impossible to skip a level in nature and produce an effect inconsonant with one's own nature. I think it would be possible for these birds' eggs, which are no larger than peas, to be laid in the pods or leaves, or to be deposited on the flowers. A flying creature might seem to be born like a flower, if the egg were broken and the seed of the bird were mixed with the moisture of the flower. Also, if a person with a vivid imagination gazes at the variety of the colours of flowers, the fantastic colours of the birds' wings might seem to be derived from the flowers. This can even be frequently seen in Europe.

Clearly, therefore, there is indeed a genuine tradition of belief, albeit one founded upon a fallacy, that small birds can be generated from flowers. Sadly, there are insufficient details to identify either the tunchon or the tunchonfung with any degree of certainty, although the latter may be a species of sunbird (nectariniid).

Native to Africa, Asia (including China's Sichuan Province), and Australasia, sunbirds are small, extremely brightly coloured, and mirror ecologically albeit not taxonomically the New World hummingbirds. Feeding primarily upon nectar, they spend much of their time in such close proximity to flowers that this intimate association may well have inspired an erroneous belief that they were actually being engendered by the flowers.

JULIUS CAESAR AND THE HERCYNIAN UNICORN

Germany has long been associated with unicorn traditions, in particular with the supposed occurrence of dainty white unicorns in the Harz Mountains. In addition, a far more exotic, bizarre version was allegedly sighted in Germany's Hercynian Forest by no less eminent an eyewitness than Julius Caesar. He described it as being an ox but shaped like a stag, the center of whose brow, between its ears, bore a single horn, taller and straighter than normal horns. Moreover, later eyewitnesses claimed that a series of branches sprouted forth from the tip of this creature's horn.

Long forgotten, the case of this very curious, atypical unicorn was

The Hercynian unicorn if indeed a uni-stag (Markus Bühler)

recently re-examined by German cryptozoologist Markus Bühler, who proposed a very ingenious, plausible explanation for it. Occasionally, a freak deer is born, bearing a single horn-like structure upon the centre of its skull, instead of its species' normal paired laterally sited antlers. (One recently recorded example is a roe deer *Capreolus capreolus* "uni-stag" born during 2007 in a park belonging to the Center of Natural Sciences in Prato, near Florence, Italy.) But what if, as speculated by Markus, some rudimentary antlers develop at the tip of this aberrant central horn?

The result would be a creature bearing a very similar appearance to the extraordinary Hercynian unicorn. So perhaps the latter beast, if truly real, was not a unicorn at all, but merely a freak uni-stag!

CHRISTOPHER COLUMBUS AND THE SKINNED SERPENT

What's in a name? Not a lot, it would seem, at least as far as cryptozoological reports are concerned. Take, for instance, the nowadays all-but-forgotten controversy surrounding the mysterious "serpent" killed by no less celebrated an explorer than Christopher Columbus, during his first voyage to the New World.

In his diary entry for October 21, 1492, Columbus reported killing and skinning a "serpent" almost 5 ft. long that he had seen entering a lake on the Bahamian island that he later dubbed Isabela. Another of these "serpents," of similar size, was killed at a different lake on that same island a day later by Martin Alonso Pinzon, captain of one of Columbus's ships, the *Pinta*.

Unfortunately, the skin of Columbus's serpent does not appear to have been preserved, which is a great pity, bearing in mind that zoologists have never been able to identify its species. Indeed, for a long time the prevailing opinion was that the creature was not a snake at all, but was instead a giant iguana. Unfortunately, however, as pointed out by Florida State Museum's assistant curator, Bill Keegan, in a fascinating *Schenectady Gazette* report (October 12, 1987) concerning this cryptozoological conundrum, the iguana identity is far from satisfactory. Iguanas do not tend to enter lakes, and there was no fossil evidence for the erstwhile existence of such creatures on this island. Also, in his diary Columbus used the word "lagartos" when referring to lizards, and "culebra" for a snake, whereas he called the beast that he and Pinzon killed a "sierpe" or serpent.

In bygone days, "serpent" had a much broader meaning than its more limited present-day usage as an alternative name for a snake. Indeed, it was often utilized as a general term for anything large and reptilian, including big snakes, lizards...and crocodiles. However, the reason why a crocodile had never been offered as a candidate for the identity of Columbus's "serpent" was that no physical evidence existed to suggest such animals had ever inhabited the Bahamas—until 1987, that is.

When Keegan led an archaeological expedition that year to Isabela, they examined the ruins of a village believed to have been visited by Columbus—and there they unearthed the left femur (thigh bone) of a crocodile. The bone was 3.5 in. long, indicating that the entire crocodile had measured around 4 ft. long. So despite the fact that there are no crocodiles here today, it would seem that in Columbus's time these reptiles did exist on Isabela—and that one such creature had suffered a grim fate at the hands of this now-famous explorer.

MEYER'S PSEUDO-PTERODACTYL—DENYING THE DRAGON

In 1696, Dutch civil engineer Cornelius Meyer published an engraving that depicted in detail the skeleton of an alleged dragon that had been obtained near Rome, Italy. The skeleton itself has long since vanished, but the engraving of it has survived, leading some cryptozoologists to speculate that what it depicts may actually have been the skeleton of a modern-day pterodactyl. If this were correct, it would be an astonishing discovery—confirming that at least one lineage of these prehistoric flying reptiles had survived into the present. The reality, however, is very different.In January 2013, a paper written by biologists Phil Senter and Pondanesa D. Wilkins from North Carolina's Fayetteville State University, and published by the scientific journal *Palaeontologia Electronica*, revealed that the skeleton of Meyer's dragon-cum-pterodactyl was a skilfully-constructed composite creation, i.e. a fake or gaff. From scrutinizing the engraving, the authors readily

Meyer's engraving of an alleged dragon-cum-pterodactyl

identified the true, and very disparate, nature of this skeleton's various components.

Namely, the skull of a domestic dog; the lower jaw of a second, smaller domestic dog; a bear's forelimb (used as the dragon's hindlimb); the ribs of a large fish; a sculpted fake tail; and a pair of fake manufactured wings. Attached portions of skin adroitly hid the junctions between these varied body parts.

Accordingly, the pterodactyls have been duly jettisoned back into 64 million years or more of Mesozoic extinction, with only Meyer's engraving remaining as silent testimony of what can result when humanity's unbounded imagination and unbridled ingenuity join forces not only to delight but also to deceive.

SONNERAT'S NON-EXISTENT PENGUINS (AND KOOKABURRA) OF NEW GUINEA

Pierre Sonnerat (1748-1814) was a French naturalist and explorer, whose publications include *Voyage à la Nouvelle-Guinée* (1776), documenting an expedition that he claimed to have made to the Spice Islands (now called the Moluccas) and New Guinea in 1771. From an ornithological standpoint, this publication is particularly intriguing, inasmuch as it reports the presence in New Guinea of no less than three species of penguin as well as the common kookaburra or laughing jackass *Dacelo novaeguineae*. His book even contains illustrations signed by him that depict the penguins as well as the kookaburra, and the brief passage in it concerning the penguins states:

> I will mention the three Manchots [penguins] which I have observed, one the Manchot of New Guinea, another the Collared Manchot of New Guinea, and the third, the Manchot Papua.

In reality, however, New Guinea is unequivocally bereft of any penguin species; and whereas three smaller kookaburra species do occur in New Guinea, the common kookaburra is confined to Australia. So how can these extraordinary discrepancies in Sonnerat's book be explained? The answer is as startling as Sonnerat's unfounded ornithological allegations.

First and foremost, Sonnerat never actually visited New Guinea! His own journey there was a complete fiction and was publicly exposed during his lifetime. Yet somehow he survived the shame with his scientific reputation intact, and the scandal was subsequently forgotten.

Conversely, his New Guinea penguins were *not* made-up birds. On the contrary, their respective species can be readily identified from their illustrations in his book. The manchot of New Guinea is the king penguin *Aptenodytes patagonicus* (breeds in northern Antarctica and various subantarctic islands); the collared manchot of New Guinea is the emperor penguin *A. forsteri* (Antarctica); and the manchot Papua is the gentoo penguin *Pygoscelis papua* (various subantartic islands including the Falklands).

Sonnerat's three New Guinea penguins: the manchot of New Guinea, left; collared manchot, centre; and manchot Papua, right (public domain)

As revealed in Penny Olsen's fascinating book, *Feather and Brush: Three Centuries of Australian Bird Art* (2001), it transpired that a number of bird skins, including those of the penguin specimens depicted in those illustrations as well as that of the kookaburra specimen depicted in its own illustration, had apparently been given to Sonnerat in 1770 at South Africa's Cape of Good Hope by English naturalist Sir Joseph Banks, who had procured them during his global travels in the 1760s. Banks instructed Sonnerat to deliver them to fellow naturalist Philibert Commerson in Mauritius. So Sonnerat sailed there, giving the skins to Commerson's draughtsman, Paul Philippe Sanguin de Jossigny, who sketched them. Following Commerson's premature death in 1773, however, Sonnerat not only kept Jossigny's illustrations of the penguins and kookaburra, but unscrupulously signed them, passing them off as his own work, and including them in his book on New Guinea.

Sadly, vestiges of Sonnerat's deception persists even today, in the misleading scientific names of the gentoo penguin and the kookaburra, which to anyone not familiar with their correct zoogeographical range suggests that the former species is native to Papua and the latter species to New Guinea.

THE GORGAKH—A DEPICTION OF DECEPTION

The gorgakh photograph (source unknown)

On August 23, 2012, I encountered a rash of online reports concerning a mysterious creature called a gorgakh that had lately been killed by villagers in Swabi, Pakistan; it was said to dig up newly buried humans from their graves and devour them. One report included the following striking photograph of unknown source, depicting the creature's carcass held upright for the camera.

This picture has also been posted on *YouTube*.

As can readily be seen from it, however, the gorgakh is merely a dead pangolin, and it appears far bigger than it actually is due to the fact that it is positioned much closer to the camera than are the various people featured in this same photograph—a classic case of forced perspective. Moreover, because pangolins are exclusively insectivorous, lack teeth, and are incapable even of chewing, we can also swiftly discount the lurid claims about its corpse-consuming proclivities.

Finally, despite some online reports inferring that pangolins are restricted to Africa, they occur in Asia too; one species, the Indian pangolin *Manis crassicaudata*, is native to parts of Pakistan. Consequently, there is no mystery as to this deceased specimen's procurement here either. Exit the gorgakh as a valid cryptid.

MASBATE SEA MONSTER—A MYSTERY NO LONGER

Not long after the Trunko investigations German cryptozoologist Markus Hemmler conducted jointly with me in September 2010 (see Chapter 8), he went on to unmask a second very notable sea monster carcass.

According to media reports, during late Christmas Eve evening/

early Christmas Day morning 1996, a grotesque-sounding carcass washed ashore on a beach near the town of Claveria in the Philippine province of Masbate. It was described in those reports as combining a turtle's head with a slender 26-ft.-long eel-like body, but it had a hole at the top of its skull that reputedly resembled a whale's blowhole. Faced with such a baffling composite, local zoologists were unable to identify its species, and it soon faded from the headlines—just another enigmatic marine carcass whose details seemed destined to be filed in the cryptozoological archives and thereafter forgotten. And except for a concise coverage included in my book *The Beasts That Hide From Man* (2003), this is precisely what happened—until October 2010, that is, when Markus, after re-examining its case, exposed its true identity.

Mindful that basking sharks *Cetorhinus maximus* possess a deceptively blowhole-like opening in their skull, which is actually the epiphysal foramen through which the pineal body extends in life, Markus checked basking shark records from the Philippines covering the time period when the Masbate carcass was beached. And sure enough, as officially documented in 2005 by shark expert Leonard Compagno and co-workers in a scientific checklist of Philippine sharks and rays (but which had not previously been accessed by the cryptozoological community), in December 1996 a basking shark had indeed been washed ashore at Masbate and formally identified. In addition, it was the first confirmed record of a basking shark from anywhere in the waters surrounding the Philippines, thus making it a significant ichthyological (albeit no longer a cryptozoological) discovery.

Success in solving this chapter's erstwhile mysteries notwithstanding, the cryptozoological archives still bulge with countless unresolved cases that had once attracted attention and interest but which have long since slipped into silent obscurity. Perhaps, as epitomized by the Trunko revelations, it is time to take another look at some of these cases, dust down their dossiers, and re-examine their stories. Who knows what we might uncover?

After all, to quote the Hungarian Nobel Prize-winning biochemist Albert Szent-Gyorgyi:

> Discovery consists in seeing what everyone else has seen and thinking what no one else has thought.

SELECTED BIBLIOGRAPHY

ANON. (1823). [Alleged discovery of living adder embedded in rock.] *The Mirror of Literature, Amusement, and Instruction*, 1 (No. 19; 8 March): 303.

ANON. (1856). [Alleged discovery of living pterodactyl embedded in rock.] *Illustrated London News*, No. 784 (9 February): 166.

ANON. (1870). A new animal to the continent [tygomelia]. *Ottawa Times* (Ottawa), 22 November.

ANON. (1883). A large turtle. *Scientific American*, 48: 292.

ANON. (1894). Monster of Issoir. *Ann Arbor Argus* (Ann Arbor), 14 September.

ANON. (1924). Fish like a polar bear. A fight with two whales. Escape after 10 days' sleep. *Daily Mail* (London), 27 December.

ANON. (1963). Is the giant lemur a "living fossil"? *New Scientist*, 20 (No. 368; 5 December): 589.

ANON. (1982). Embeddings. *Fortean Times*, No. 36 (Winter): 17-19.

ANON. (1988). Another Siberian find. *Fortean Times*, No. 51 (Winter): 11.

ANON. (1993). Are there Nessies in the Pacific? *Ceefax* (London), 30 September.

ANON. (1997). Sea creature a mystery [Masbate sea monster carcass]. *South China Morning Post* (Hong Kong), 25 February.

ANON. (2005). Vet: Dead rabbit looks like a jackalope. *Boston Globe* (Boston), 29 August.

ANON. (2009). The legend of Trunco [sic]. *Margate Business Association*, http://www.margatebusiness.co.za/index.php?option=com_content&view=article&id=64, 19 October.

ANON. (2011). World's heaviest spider title challenged at Museum. Natural History Museum website, http://www.nhm.ac.uk/about-us/news/2011/july/worlds-heaviest-spider-title-challenged-at-museum99065.html, 15 July.

ANON. (2013). Angolan witch spider. *Snopes.com*, http://www.snopes.com/photos/bugs/witchspider.asp, 8 February.

ALLEN, Andrew (1984). Toad in the prehistoric hole. *The Countryman*, 89 (Spring): 62-64.

ANNANDALE, N. (1908). An unknown lemur from the Lushai

Hills, Assam. *Proceedings of the Zoological Society of London*, (17 November): 888-889.

BARON, R[ichard]. (1890). A Malagasy forest. *Antananarivo Annual and Madagascar Magazine*, 4: 196-211.

BASSETT, Michael G. (1982). *"Formed Stones," Folklore and Fossils.* National Museum of Wales (Cardiff).

BAYLESS, Mark K. (2000). Giant lizards, salamanders, snakes, and turtles. *Fate*, 53 (November): 25-27.

BECK, Jane C. (1972). The giant beaver: a prehistoric memory? *Ethnohistory*, 19 (Spring): 109-122.

BONDESON, Jan (2007). Toad in the hole. *Fortean Times*, No. 221 (April): 38-42.

BRADSHAW, Lance (n.d.). What about Trunko? *Kryptid's Keep*, http://www.angelfire.com/sc2/Trunko/trunko.html, accessed 15 January 2007.

BULPIN, T.V. (1965). *Your Undiscovered Country.* T.V. Bulpin (Cape Town).

BURNEY, David A. & RAMILISONINA (1998). The *kilopilopitsofy, kidoky,* and *bokyboky*: Accounts of strange animals from Belo-sur-mer, Madagascar, and the megafaunal "extinction window." *American Anthropologist*, 100 (No. 4; December): 957-966.

BURTON, Maurice (1960). The Soay beast. *Illustrated London News*, 236 (4 June): 972-973.

BURTON, Maurice (1961). Was the Soay beast a tourist? *Illustrated London News*, 239 (14 October): 632.

CASTLEDEN, Rodney (1987). An occasional hell: The mystery of the entombed toads. *The Unknown*, No. 24 (June): 14-18.

COLEMAN, Loren. (2010). Camel spiders and other alleged giant spiders. *Cryptomundo*, http://www.cryptomundo.com/cryptozoo-news/giantspiders-4, 11 November.

COLEMAN, Loren & HUYGHE, Patrick (2003). *The Field Guide to Lake Monsters, Sea Serpents, and Other Mystery Denizens of the Deep.* Tarcher/Penguin (New York).

COOPER, John (1993). Frogs alive. *New Scientist*, 138 (8 May).

CORLISS, William M. (1980). *Unknown Earth: A Handbook of Geological Enigmas.* The Sourcebook Project (Glen Arm).

COUDRAY, Philippe (2009). *Guide des Animaux Cachés.* Éditions Du Mont (Cazouls-les-Béziers).

Selected Bibliography

CUNINGHAME, R.J. (1912). The water-elephant. *Journal of the East Africa and Uganda Natural History Society*, 2: 97-99.

DANCE, S. Peter (1976). *Animal Fakes and Frauds.* Sampson Low (Maidenhead).

DASH, Mike (2010). Baron Von Forstner and the U28 sea serpent of 1915. *A Fortean in the Archives,* http://aforteantinthearchives.wordpress.com/2010/01/08/baron-von-forstner-and-the-u28-sea-serpent-of-1915/, 8 January.

DAVIES, Steve (2003). *Fossils in Lyme Regis.* Dinosaurland Fossil Museum (Lyme Regis).

DAWES, Colin (2003). *Fossil Hunting Around Lyme Regis.* Colin Dawes Studios (Lyme Regis).

DURRELL, Gerald (1954). *The Bafut Beagles.* Rupert Hart-Davis (London).

EBERHART, George M. (2002). *Mysterious Creatures: A Guide to Cryptozoology* (2 vols). ABC-Clio (Santa Barbara).

EDWARDS, Frank (1959). *Stranger Than Science.* Ace Books (New York).

FALCONI, Marta (2008). Single-horned "unicorn" is deer found in Italy. Associated Press (London), 14 June.

FAWCETT, Percy H. (1953). *Exploration Fawcett.* Hutchinson (London).

FINN, R. Anthony (1982). Horned hares. *Shooting Times*, No. 4201 (9 September): 92.

FLACOURT, Étienne de (1658). *L'Histoire de la Grand Îsle de Madagascar.* Gervais Clousier (Paris).

FORT, Charles H. (1931). *Lo!* Charles Kendall (London).

FOWLER, Terence (2006). Entombed amphibian. *Fortean Times*, No. 206 (February): 75.

GAZZOLA, Alex (1997). Toads entombed. *Enigma*, No. 5 (Autumn): 43-44.

GIBBONS, William J. (2001). Giant spiders. cz@yahoogroups.com, 21 October.

GIBBONS, William J. (2004). A Dictionary of Cryptozoology. cryptolist@yahoogroups.com, 24 May.

GIBBONS, William J. (2004). Note on giant spider. cryptolist@yahoogroups.com, 11 June.

GIBBONS, William J. (2006). *Missionaries and Monsters.* Coachwhip Publications (Landisville).

GIBBONS, William J. & HOVIND, Kent (1999). *Claws, Jaws, and Dinosaurs.* CSE Publications (Pensacola).

GOODMAN, Steven M. (1994). Description of a new species of subfossil eagle from Madagascar: *Stephanoaetus* (Aves: Falconiformes) from the deposits of Ampasambazimba. *Proceedings of the Biological Society of Washington*, 107: 421–428.

GREENWELL, J. Richard (Ed.) (1986). Hippoturtleox. *ISC Newsletter*, 5 (No. 1; Spring): 10.

HAND, Suzanne & ARCHER, Michael (Eds) (1987). *The Antipodean Ark.* Angus & Robertson (North Ryde).

HEALD, Daniel (1993). Mystery frog. *New Scientist*, 138 (10 April).

HEMMLER, Markus (2010). Trunko-Recherche: Übersicht des Aktuellen Stands. *Kryptozoologie-Online*, http://www.kryptozoologie-online.de/Nachrichten/Dracontologie-News/trunko-recherche-uebersicht-des-aktuellen-stands.html, 10 September.

HEUVELMANS, Bernard (1958). *On the Track of Unknown Animals.* Rupert Hart-Davis (London).

HEUVELMANS, Bernard (1968). *In the Wake of the Sea-Serpents.* Rupert Hart-Davis (London).

HEUVELMANS, Bernard (1978). *Les Derniers Dragons d'Afrique.* Plon (Paris).

HEUVELMANS, Bernard (1986). Annotated checklist of apparently unknown animals with which Cryptozoology is concerned. *Cryptozoology*, 5: 1-26.

HICHENS, William (1937). African mystery beasts. *Discovery*, 18 (December): 369-373.

HOLLIDAY, Chuck & JAPUNTICH, Dan (2010). Jackalopes. http://sites.lafayette.edu/hollidac/links-for-fun/jackalopes/, 6 April.

HOWE, S.R. *et al.* (1981). *Ichthyosaurs: A History of Fossil "Sea-Dragons."* National Museum of Wales (Cardiff).

HUGHES, Gareth (1995). The amazing story of frog-in-the-Coke-hole. *Daily Post* (North Wales), 27 October.

HYNES, Peter (2001). Giant spiders. cz@yahoogroups.com, 20 October.

JONES, A.K. (1925) [Description and photographs of the Trunko carcass]. *In*: ANON. (1925). The case for the sea-serpent. A remarkable batch of "sequels." *Wide World Magazine*, 55 (No. 2; August): 304-5.

Selected Bibliography 169

KARATZAS, Valerie (2007). Toad in the hole. *Fortean Times*, No. 226 (August): 75.

KIRCHER, Athanasius (1667). *China Illustrata*. Jacobum (Meurs).

LAMBERT, Dixie (2008). "Sea monster" discovery on Glacier Island the buzz of old Cordova. *Cordova Times* (Cordova), 2 May.

LAVAUDEN, Louis (1931). Animaux disparus et légendaires de Madagascar. *Revue Scientifique*, 10: 297-308.

MACKAL, Roy P. (1987). *A Living Dinosaur? In Search of Mokele-Mbembe*. E.J. Brill (Leiden).

MAGIN, Ulrich (1988). Forstner sea serpent sighting: A possible hoax? *Strange Magazine*, No. 2: 4.

MAREŠ, Jaroslav (1995). *Detektivem v Říši Zvířat*. Magnet Press (Prague).

MARTIN, Douglas (2003). Douglas Herrick, 82, dies; father of West's jackalope. *New York Times* (New York), 19 January.

McGOWAN, Christopher (1991). *Dinosaurs, Spitfires, and Sea Dragons*. Harvard University Press (Cambridge).

MEHR, Michael (2007). Beastly nationalism. *Fortean Times*, No. 228 (October): 57.

MICHELL, John & RICKARD, Robert J.M. (1982). *Living Wonders*. Thames & Hudson (London).

MILLER, Penny (1979). *Myths and Legends of Southern Africa*. T.V. Bulpin (Cape Town).

NEWTON, Michael (2005). *Encyclopedia of Cryptozoology: A Global Guide*. McFarland (Jefferson).

NEWTON, Michael (2012). *Globsters*. CFZ Press (Bideford).

NIEWHOFF [aka NIUHOFF, NEUHOFF], Johan (1669). *Die Gesantschaft der Ost-Indischen Gesellschaft in den Vereinigten Niederländern*. Jacob Mörs (Amsterdam).

OLSEN, Penny (2001). *Feather and Brush: Three Centuries of Australian Bird Art*. CSIRO (Melbourne).

PARÉ, Ambroise (1982). *Of Monsters and Marvels* [based upon the Malgaigne edition of 1840; translated with an introduction and notes by Janis L. Pallister]. University of Chicago Press (Chicago).

PETIT, Georges (1930). *L'Industrie des Pêches à Madagascar*. Société d'Éditions Géographiques, Maritimes et Coloniales (Paris).

POCOCK, Reginald I. (1899). On the scorpions, pedipalps and spiders from tropical West- Africa, represented in the collection of the

British Museum. *Proceedings of the Zoological Society of London*, (14 November): 833-885 [p. 844].

PRITCHARD, Peter C.H. (2012). *Rafetus, the Curve of Extinction: The Story of the Giant Softshell Turtle of the Yangtze and Red Rivers.* Living Art Publishing (Ada).

PURCELL, Rosamond W. & GOULD, Stephen J. (1992). *Finders, Keepers: Eight Collectors.* Hutchinson Radius (London).

QUESADA, Carlos Nores & VONLETTOW-VORBECK, Corina Liesau (1992). La zoologia historica como complemento de la arqueozoologia. El caso del zebro. *Archaeofauna*, 1: 61-71.

ROSE, Tom (2012). Giant alien "gorgakh" killed in Pakistan (video). *Before It's News*, http://beforeitsnews.com/paranormal/2012/08/giant-alien-gorgakh-killed-in-pakistan-video-2442784.html, 23 August.

SCHOMBURGK, Hans (1957). *Zelte in Afrika.* Verlag der Nation (Leinen).

SCREETON, Paul (1983). The enigma of entombed toads. *Fortean Times*, No. 39 (Spring): 36-39.

SEDGWICK, Paulita (1974). *Mythological Creatures: A Pictorial Dictionary.* Holt, Rinehart & Winston (New York).

SENTER, Phil & WILKINS, Pondanesa D. (2013). Investigation of a claim of a late-surviving pterosaur and exposure of a taxidermic hoax: the case of Cornelius Meyer's dragon. *Palaeontologia Electronica*, http://palaeo-electronica.org/content/2013/384-late-surviving-pterosaur, 19 January.

SHUKER, Karl P.N. (1995). *In Search of Prehistoric Survivors: Do Giant 'Extinct' Creatures Still Exist?* Blandford (London).

SHUKER, Karl P.N. (1999). *Mysteries of Planet Earth: An Encyclopedia of the Inexplicable.* Carlton (London).

SHUKER, Karl P.N. (2003). *The Beasts That Hide From Man: Seeking the World's Last Undiscovered Animals.* Paraview Press (New York).

SHUKER, Karl P.N. (2007). *Extraordinary Animals Revisited: From Singing Dogs To Serpent Kings.* CFZ Press (Bideford).

SHUKER, Karl P.N. (2008). *Dr Shuker's Casebook: In Pursuit of Marvels and Mysteries.* CFZ Press (Bideford).

SHUKER, Karl P.N. (2010). Behold, Trunko!! *ShukerNature*, http://karlshuker.blogspot.com/2010/09/behold-trunko.html 5 September.

Selected Bibliography

SHUKER, Karl P.N. (2010). Trunko—two more photographs!! *ShukerNature*, http://karlshuker.blogspot.com/2010/09/trunko-two-more-photographs.html, 9 September.

SHUKER, Karl P.N. (2010). Son of Trunko! *ShukerNature*, http://karlshuker.blogspot.com/2010/09/son-of-trunko.html, 16 September.

SHUKER, Karl P.N. (2010). Trunko—solved after 85 years! [in Alien Zoo]. *Fortean Times*, No. 268 (November): 19.

SHUKER, Karl P.N. (2010). *Karl Shuker's Alien Zoo: From the Pages of Fortean Times.* CFZ Press (Bideford).

SHUKER, Karl P.N. (2011). Trunko—a trio of world-exclusives on *ShukerNature.* Here's how it happened! *In*: DOWNES, Corinna & DOWNES, Jonathan (Eds) (2011), *The CFZ Yearbook 2011.* CFZ Press (Bideford): 7-20.

SHUKER, Karl P.N. (2011). Trunko returns!! A fourth photograph is discovered! *ShukerNature*, http://karlshuker.blogspot.co.uk/2011/03/trunko-returns-fourth-photograph-has.html, 10 March.

SHUKER, Karl P.N. (2012). *The Encyclopaedia of New and Rediscovered Animals: From The Lost Ark To The New Zoo—And Beyond.* Coachwhip Publications (Landisville).

SIBREE, James (1889). The oratory, songs, legends, and folk-tales of the Malagasy. *Antananarivo Annual and Madagascar Magazine*, 14: 364-368 [relevant section].

SKINNER, Bob (1985). *Toad in the Hole.* Fortean Times (London).

SKINNER, Bob (1987). A wonder from the tundra. *Fortean Times*, No. 49 (Winter): 4-5.

"SKINNERBOY" [=SKINNER, James]. (2012). Kotoko: Madagascar's Little Hairy Men. *YouTube* video, http://www.youtube.com/watch?v=VCh9Yf4E81g&, uploaded 18 November.

SLÁMA, Vojtìch (2002). *Hon na Vodní Pøíšery.* Regia (Prague).

SMITH, G. Herbert (1886). Some Betsimisaraka superstitions. *Antananarivo Annual and Madagascar Magazine*, 10: 239-243.

SPLITTER, Henry W. (1954). The impossible fossils. *Fate*, 7 (January): 65-72.

STAN, Alexandru (2008). In search for the 8-legged freak. *InOut Star*, http://www.inoutstar.com/news/In-Search-For-The-8-Legged-Freak-5443.html, 12 March.

STELLER, Georg W. (1751). *De Bestiis Marinis*. Reprinted 2011 by Nabu Press (USA).

SUTTON, Jane (1987). Columbus' "serpent" probably a croc, says archeologist who found bone. *Schenectady Gazette* (Schenectady), 12 October,

TEGNER, Henry (1975). Man made the unicorn. *The Countryman*, 80 (No. 2; Summer): 182-184.

TSCHCHIKVADSE, V.M. & TERBISCH, C. (1989). Das Geheimnis der steinernem Schildkröten. *Naturwissenschaftliche Rundschau*, 42 (No. 3): 110-11.

VAŠÍČEK, Arnošt (1998). *Planeta Záhad: Tajemná Minulost*. Baronet (Prague).

VENABLES, Tony (2011). Trunko—probably. *Fortean Times*, No. 277 (July): 68-69.

WHALEN, Dwight (1988). Stonewalled bat. *Fate*, 41 (May): 71.

WHITEHEAD, G.K. (1982). Chimerical hare. *Shooting Times*, No. 4206 (14 October): 19.

WILKINSON, J.R. (1993). Dried frogs. *New Scientist*, (15 May).

WILSON, Jane (1990). *Lemurs of the Lost World*. Impact Books (London).

WOODRUFF, Una (1979). *Inventorum Natura: The Expedition Journal of Pliny the Elder*. Dragon's World (London).

WOODRUFF, Una (1981). *Amarant: The Flora and Fauna of Atlantis by a Lady Botanist*. Dragon's World (London).

WOOLHEATER, Craig (2013). More on camel spiders and other alleged giant spiders. *Cryptomundo*, http://www.cryptomundo.com/cryptotourism/giantspiders-3/, 8 April.

ACKNOWLEDGEMENTS

Many people have assisted my endeavors in researching and writing this book, but I wish to offer my special thanks to my editor/publisher Patrick Huyghe for his interest and enthusiasm concerning it, to Anthony Wallis for his truly spectacular front-cover artwork, and also to the following persons, publications, and organizations for their varied and much-valued contributions:

David Alderton, Bianca Baldi, George Beccaloni and Jan Beccaloni, Adam Bird, Lance Bradshaw, Markus Bühler, Centre for Fortean Zoology, Loren Coleman, *The Dorsetarian*, Jonathan Downes, Miroslav Fišmeister, *Flying Snake*, *Fortean Times*, Ray Gabriel, Marc Gaglione, Richard Gallon, Bill Gibbons, Craig Glenday, Mike Grayson, *Guinness World Records*, Markus Hemmler, David Heppell, the late Bernard Heuvelmans, Richard Holland, Peter Horler, Roy P. Mackal, the late Ivan Mackerle, Ulrich Magin, Ralph Molnar, Rob Morphy, Tim Morris, Richard Muirhead, Darren Naish, Natural History Museum London, Michael Newton, Mark North, *Paranormal*, *Practical Reptile Keeping*, Michel Raynal, William M. Rebsamen, Maria E. Rutzmoser, Robert Schneck, the late Mary D. Shuker, Bob Skinner, James Skinner, *Small Furry Pets*, Kevin Stewart, Richard Svensson, Lars Thomas, Gerard Van Leusden, John Warms, Richard Wells.

ABOUT THE AUTHOR

Born and still living in the West Midlands, England, Karl P.N. Shuker graduated from the University of Leeds with a Bachelor of Science (Honors) degree in pure zoology, and from the University of Birmingham with a Doctor of Philosophy degree in zoology and comparative physiology. He now works full-time as a freelance zoological consultant to the media, and as a prolific published writer.

Shuker is currently the author of 19 books and hundreds of articles, principally on animal-related subjects, with an especial interest in cryptozoology and animal mythology, on which he is an internationally recognized authority, but also including a poetry volume. In addition, he has acted as consultant for several major multi-contributor volumes as well as for the world-renowned *Guinness Book of Records/Guinness World Records* (he is currently its Senior Consultant for its Life Sciences section); and he has compiled questions for the BBC's long-running cerebral quiz *Mastermind*. He is also the editor of the *Journal of Cryptozoology*, the world's only existing peer-reviewed scientific journal devoted to mystery animals.

Shuker has travelled the world in the course of his researches and writings, and has appeared regularly on television and radio. Aside from work, his diverse range of interests include motorbikes, the life and career of James Dean, collecting masquerade and carnival masks, quizzes, philately, poetry, travel, world mythology, and the history of animation.

He is a Scientific Fellow of the prestigious Zoological Society of London, and a Fellow of the Royal Entomological Society. He is Cryptozoology Consultant to the Centre for Fortean Zoology, and is also a Member of the Society of Authors.

His personal website can be accessed at http://www.karlshuker.com and his mystery animals blog, ShukerNature, can be accessed at http://www.karlshuker.blogspot.com

His poetry blog can be accessed at http://starsteeds.blogspot.com and his Eclectarium blog can be accessed at http://eclectariumshuker.blogspot.com

There is also an entry for Karl Shuker in the online encyclopedia Wikipedia at http://en.wikipedia.org/wiki/Karl_Shuker and a Like (fan) page on Facebook.

AUTHOR BIBLIOGRAPHY

Mystery Cats of the World: From Blue Tigers To Exmoor Beasts (Robert Hale: London, 1989)

Extraordinary Animals Worldwide (Robert Hale: London, 1991)

The Lost Ark: New and Rediscovered Animals of the 20th Century (HarperCollins: London, 1993)

Dragons: A Natural History (Aurum: London/Simon & Schuster: New York, 1995; republished Taschen: Cologne, 2006)

In Search of Prehistoric Survivors: Do Giant "Extinct" Creatures Still Exist? (Blandford: London, 1995)

The Unexplained: An Illustrated Guide to the World's Natural and Paranormal Mysteries (Carlton: London/JG Press: North Dighton, 1996; republished Carlton: London, 2002)

From Flying Toads To Snakes With Wings: From the Pages of FATE Magazine (Llewellyn: St Paul, 1997; republished Bounty: London, 2005)

Mysteries of Planet Earth: An Encyclopedia of the Inexplicable (Carlton: London, 1999)

The Hidden Powers of Animals: Uncovering the Secrets of Nature (Reader's Digest: Pleasantville/Marshall Editions: London, 2001)

The New Zoo: New and Rediscovered Animals of the Twentieth Century [fully-updated, greatly-expanded, second edition of *The Lost Ark*] (House of Stratus Ltd: Thirsk, UK/House of Stratus Inc: Poughkeepsie, USA, 2002)

The Beasts That Hide From Man: Seeking the World's Last Undiscovered Animals (Paraview Press: New York, 2003)

Extraordinary Animals Revisited: From Singing Dogs To Serpent Kings (CFZ Press: Bideford, 2007)

Dr Shuker's Casebook: In Pursuit of Marvels and Mysteries (CFZ Press:

Bideford, 2008)

Dinosaurs and Other Prehistoric Animals on Stamps: A Worldwide Catalogue (CFZ Press: Bideford, 2008)

Star Steeds and Other Dreams: The Collected Poems (CFZ Press: Bideford, 2009)

Karl Shuker's Alien Zoo: From the Pages of Fortean Times (CFZ Press: Bideford, 2010)

The Encyclopaedia of New and Rediscovered Animals: From The Lost Ark to The New Zoo—and Beyond [fully-updated, greatly-expanded, third edition of *The Lost Ark*] (Coachwhip Publications: Landisville, 2012)

Cats of Magic, Mythology, and Mystery: A Feline Phantasmagoria (CFZ Press: Bideford, 2012)

Mirabilis: A Carnival of Cryptozoology and Unnatural History (Anomalist Books: New York, 2013)

Consultant and also Contributor

Man and Beast (Reader's Digest: Pleasantville, New York, 1993)

Secrets of the Natural World (Reader's Digest: Pleasantville, New York, 1993)

Almanac of the Uncanny (Reader's Digest: Surry Hills, Australia, 1995)

The Guinness Book of Records/Guinness World Records 1998-present day (Guinness: London, 1997-present day)

Consultant

Monsters (Lorenz: London, 2001)

Contributor

Of Monsters and Miracles CD-ROM (Croydon Museum/Interactive Designs: Oxton, 1995)

Fortean Times Weird Year 1996 (John Brown Publishing: London,

1996)

Mysteries of the Deep (Llewellyn: St Paul, 1998)

Guinness Amazing Future (Guinness: London, 1999)

The Earth (Channel 4 Books: London, 2000)

Mysteries and Monsters of the Sea (Gramercy: New York, 2001)

Chambers Dictionary of the Unexplained (Chambers: Edinburgh, 2007)

Chambers Myths and Mysteries (Chambers: Edinburgh, 2008)

The Fortean Times Paranormal Handbook (Dennis Publishing: London, 2009)

Plus numerous contributions to the annual *CFZ Yearbook* series of volumes.

Editor

Journal of Cryptozoology (published by CFZ Press)

INDEX OF ANIMALS

Italic page numbers refer to illustrations.

Adder, entombed, 119-120
Aepyornis maximus, 89, 129
Al-mi'raj, 31
Alces alces, 50
Aldabrachelys arnoldi, 19
  gigantea, 18-19
  grandidieri, 91
  hololissa, 19
Ammonite, 63, 66-68, *67*, 120
Amphichelydid, 22
Anomalurus derbianus orientalis, 152
Antamba, 85-86
Antilocapra americana, 31
Aptenodytes forsteri, 161
  patagonicus, 161
Archaeolemur, 76
Archelon ischyros, 11
Arietites, *67*, 67
Ascaris lineata, 121
Ass, European wild, 54
Aye-aye, 79
  giant, 79

Babirusa, 8-9, *9*
Babyrousa, 8
Balaenoptera acutorostrata, 106
Basilisk, 118, 120-121, 156
Bassaricyon, 61
Bat, scarlet tree-entombed, 122-123
Bear, polar, 50-52, 95, 97, 99
Beaver, American, *142*
  giant, 1, 141-146, *142*
  northern giant, 145
  southern giant, 145
Belemnite, 68-69, *69*

Bird, flower-generated, 155
Bixi, 22
Blue Ben of Kilve, 64
Brachyplatystoma filamentosum, 60
Busco beast, 24-25

Capercaillie, 31
Capreolus capreolus, 158
Castor canadensis, 141
Castoroides leiseyorum,
  ohioensis, *144*, 144, 145
Catfish, European giant, 56
Cetorhinus maximus, 163
Chelonia mydas, 15
Chelonoidis nigra, 16
  nigra abingdoni, 16, 18
  nigra nigra, 18
  nigra phantastica, 18
Chipekwe, 147-148
Choeropsis liberiensis, 80, 146, 148
Colossochelys atlas, 12
Con rit, Vietnamese, 87
Coroniceras, 67
Crocodile, dwarf, 125-127
  estuarine (=saltwater), 129-130, 133-134, 137
  fossil, 66
  Johnston's (=freshwater), 134
  Madagascan horned, 128
  Mary River mystery, 134
  Murua, 135
  Nile, 125-128
  sea, 134, 135-138
  tailless, 130-132
Crocodile-frog, giant Bornean, 129-132, *130*
Crocodylus johnstoni, 134
  niloticus, 126

porosus, 129
Cryptoprocta ferox, 85
   spelea, 85
Cynocephalus, 88

Dacelo novaeguineae, 160
Daubentonia madagascariensis, 179
   robusta, 91
Deer, roe, 29, 158
Deinosuchus (=*Phobosuchus*), 127
Dermochelys coriacea, 11
Devil's toenail, 68-69
Didelphis virginiana, 61
Dimorphodon, 64
Dipsochelys dussumieri, see *Aldabrachelys
gigantea,* 18
Dragon, 1,2, 22, 29, 63, 64, 66, 69, 132-
133, *133*, 156, *159*, 159, 160
     Meyer's, *159*, 159-160
     Nepalese crocodile, *133*
     sea, 1, 64-66, *65*, 69

Eagle, Malagasy crowned, 90, 147
Elephant, water, 56-58, *57*
Elephant bird, great, 71, 88-90, 129
Equus hemionus, 54
   hydruntinus, 54-55

Fairy loaf, 68-69
Fangalabolo, 91
Father-of-All-the-Turtles, 11-12
Felis silvestris lybica, 91
Fish, pearl-embedded, 118-119
Flying jackal, Tanzanian, *151*, 152
Flying mouse, *149*, 149-151
     Kivu, 150
     long-eared, 149-151
     Zenker's, 149
Fossa, 85-86

giant, 85
melanistic, 86
Fotsiaondré, 77
Frog, entombed, 111-114
   entombed tree, 113
   water-holding, *117*, 117

Gavialis gangeticus, 133
   papuensis, 135
Gharial (=Gavial), 125, 133-135
Glacier Island sea monster, 106
Globster, 103-109
Gorgakh, *162*, 162
Gryphaea arcuata, 69

Habéby, 77-78
Hadropithecus, 72, 75, 76, 81
Hare, horned, 2, *29*, 29-30, 32, 33
     unicorn, 29, 31, 53, 156, *157*,
     157-158
Hexaprotodon liberiensis, 80
   madagascariensis, 79
Hippopotamus, common, 79, 146
   lesser Malagasy, 79
   Malagasy dwarf, 79-80
   Malagasy pygmy, 79
   pygmy, 79-80, 146
Hippopotamus amphibius, 79, 146
   laloumena, 79
   lemerlei, 79-80
Hippoturtleox, 2, 25-26
Hololissa, 19
Hound, wild American, 60-61, *61*
Hydrodamalis gigas, 50
Hyla, 113
Hynobius, 118
Hysterocrates hercules, 41-42

Ichthyosaur, 63-66, *65*

Index of Animals

Idiurus kivuensis, 149-150
    macrotis, 149-151
    zenkeri, 149-150
Indri, 76, 81, 88
Indri indri, 76, 88

J'ba fofi, 37, 47
Jackalope, 2, *31*, 31-33

Kalanoro (=Kolanaro), 80-85
Kidoky, 76, 79
Kilopilopitsofy, 79-80
Kinkajou, 61
Koo-be-eng, 147
Kookaburra, common, 160-162
    New Guinea common, 160-162
Kotoko, 82-85, *83*
Kou-teign-koo-rou, 147

Lake Duobuzhe monster, 2, 25-26
Lemur, giant, 2, 71-80, 90-91, 128
    koala, *75*, 75-76
    sloth, 38, 72-73, *73*
Lepus cornutus, 30
Lipata, 126-127
Litoria platycephala, 117
Lonesome George, 16-18
Lord-of-the-Sea, Madagascan, 86
Loris, Indian slow, 58-59
    tailed slow, *58*, 58-59

Macroclemys temminckii, 25
Mahamba, 125-128, *126*
Malagnira, 91
Man, dog-headed, 87-88
Mangarsahoc, 78-79
Manis crassicaudata, 162
Margaritifera margaritifera, 119
Masbate sea monster, 153, 162-163

Mastodonsaurus, 131
Megaladapis edwardsi, *75*, 75, 77-78
Megalopedus, tusked, 3-5, *4*, 9
Meiolania brevicollis, 26
    damelipi, 26
    mackayi, 26
    platyceps, 26
Mekosuchine, 131
Mekosuchus inexpectatus, 131
Micraster, 68
Mlularuka, *151*, 151-152
Moeritherium, 57
Mokele-mbembe, 11, 22, 37, 125, 147
Moose, 2, 49, 50
Mosasaur, 137
Moth, mystery Madagascan long-tongued, 91
Mouse, throat-blocking, 122
Mussel, common pearl, 119

Ndendecki, *23*, 23-24, 25
Ndgoko na maiji, 57
Newt, permafrost-entombed, 117-118
Nigbwe, 146
Nudibranch, 153-154
Nycticebus bengalensis, 59
    caudatus, 59

Olingo, 61
Onager, 54
Opossum, Virginia, 61
Osteolaemus, 125, 127

Pa beuk, 60
Palaeopropithecus ingens, 72-73, *73*
Pangasianodon gigas, 60
Pangolin, 162
    Indian, 162
Pelinobius muticus (=*Citharischius crawshayi*), 41

Pelochelys bibroni, 27
Penguin, 160-162
 emperor, 161
 gentoo, 161
 king, 161
 New Guinea, 2, 161, *161*
Piraiba, 60
Plesiosaur, 63, 64, *65*, 137
Pliosaur, 137
Potos flavus, 61
Pronghorn, 31-32
Pterodactyl, 1, 2, 115, *159*, 159-160
 entombed, 115
Pterodactylus anas, 115
Pterosaur, 64
Pygoscelis papua, *161*, 161

Rabbit, cottontail, 32
 jack, 31
Rabbit-bird, 30
Rafetus leloii, 27
 swinhoei, 27
Railalomena, 91, 128
Rhamphosuchus, 127
Roc (=Rukh), *88*, 88-90,
Roundworm, ascarid, 121

Salamander, ice-entombed Asian, 118
Sarcosuchus, 127
Scaly-tail(=Anomalure), 149
 Lord Derby's (=Fraser's), 152
 pygmy, 149
Sea centipede, 87
Sea millipede, 87
Sea serpent, 1, 12, 66, 86, 87, 97, 109, 135-139, 136
 Grangense, 135, *136*
 Kompira Maru, 136
 Sacramento, 136

U-28, 138, *139*
Sea urchin, 68
Sea-ape, Steller's, 51
Sea-bear, Steller's, 51-52
Sea-cow, Steller's, 50, 52
Sea-raven, Steller's, 51
Seal, fur, 51
Serpent, Christopher Columbus's mystery giant, 2, 158-159
 stone, 66-67
Shark, basking, 109, 163
 Paraguay River toothless, 59
Sifaka, 75, 76
Silurus glanis, 56
Skvader, 31
Slal'i'kum, 142, 143
Slug, sea, 53
Snail, antlered Sarmatian sea, 2, 52-53, *53*
Soay beast, 14-15
Softshell, African, 24
 Chinese, 27
 New Guinea giant, 27
Solifugid, 44
Sorraia, 55
Spider, American giant mystery, 43
 Angolan witch, 44
 British giant mystery, 43-44
 French giant mystery singing, 44-46
 goliath bird-eating, 35, 40
 hercules baboon, 40-43
 king baboon, 41
 New Guinea giant mystery, 36
 pinkfoot goliath bird-eating, 35
 South American giant mystery, 38-39
 Vietnamese giant mystery, 39
Squid, giant, 96, 101, 105
Squirrel, Schomburgk's Liberian micro-, 2, 146-151
Stegodon florensis, 9

Index of Animals   185

Stegodont, 9
Stephanoaetus mahery, 90
Sturgeon, 60
Sukotyro, 2, 5-9, *7*
Sukotyro indicus, 7

Temnospondyl, 131
Tetrao urogallus, 31
Thalattosuchian, 134-135, 137
Thanacth, 72, *72*
Theraphosa apophysis, 35
    blondi, 35, 40, *40*, 42-43
Thunder bullet, 68, 69
Titanites, 67
Toad, entombed, 111-115, *112*, 118
   throat-blocking, 122
Tokandia, 75-76
Tompondrano, 86-87
Tortoise, Aldabra giant, 18-19
   Arnold's giant, 19
   Fernandina (=Narborough) Island
giant, 18
   Floreana (=Charles) Island giant, 18
   Galapagos giant, 16-18, *17*
   Madagascan giant, 19
   Paracels giant, 20
   Pinta (=Abingdon) Island giant, 16-18, *17*
Tratratratra, 71-75, 78
Trionyx triunguis, 24, 25
Trunko, 2, *93*, 93-109, *102, 105, 107*, 163
Tsomgomby, 79-80
Tu, 147, 148
Tunchonfung, 156-157
Turtle, alligator snapping, 25
   American snapping, 21
   Annie L. Hall giant mystery, 13
   Christopher Columbus's giant mystery, 13
   flying, 2, 153-155, *154*, 156
   Hoan Kiem giant freshwater, 27-28

Indian Ocean giant mystery, 12, 18-19
leathery (=leatherback), 11, 13, 15
leucistic black (=green) sea, 15
Rhapsody giant mystery, 13, 15
Sri Lankan giant mystery, 12-13
yellow giant mystery, 13, 15
Tygomelia, 2, 49-50, *50*

Uni-stag, *157*, 158
Unicorn, 2, 29, 31, 53, 156, 157, 158
   Harz Mountains, 157
   Hercynian, 2, *157*, 157-158
Ursus maritimus, 51

Vipera berus, 119
Voay robustus, 128
Vorompatra (=vouronpatra), 89-90

Water-leopard, 147
Wels, 56, 60
Whale, minke, 106
Whale-fish, Lake Myllesjön, 55-56
Wildcat, African, 91
   Madagascan mystery, 91
Wolpertinger, *30*, 30-31

Zebra, 55
Zebro (=Encebro), 54-55

CPSIA information can be obtained at www.ICGtesting.com
Printed in the USA
BVOW07s1059060514

352713BV00011B/580/P